ONE GIANT LEAP

Iconic and Inspiring Space Race Inventions
that Shaped History

CHARLES PAPPAS

LYONS
PRESS

Guilford, Connecticut

An imprint of The Rowman & Littlefield Publishing Group, Inc.
4501 Forbes Blvd., Ste. 200
Lanham, MD 20706
www.rowman.com

Distributed by NATIONAL BOOK NETWORK

British Library Cataloguing in Publication Information available

Library of Congress Cataloging-in-Publication Data available

ISBN 978-1-4930-3843-5 (hardcover)
ISBN 978-1-4930-3844-2 (e-book)

∞™ The paper used in this publication meets the minimum requirements of American National Standard for Information Sciences—Permanence of Paper for Printed Library Materials, ANSI/NISO Z39.48-1992.

Printed in the United States of America

*Dedicated to everyone who agrees with Larry Niven that
"The dinosaurs became extinct because they didn't have a space program."*

CONTENTS

FOREWORD

AT THIS FIFTIETH ANNIVERSARY OF APOLLO 11, THE FIRST HUMAN LAND-
ing on the Moon, many Americans will reflect on an epoch-defining event.
Some will nostalgically commemorate an incredible national achievement—*twelve Americans* trod the lunar surface from 1969-72—that
demonstrated America's pluck and verve.

The Apollo moon program elicits both nostalgia and skepticism
because it dramatically culminated an era of barnstorming astronautics
that seemingly failed to transform the arc of human history. But people
who would measure that transformation only in atomic-powered bases
in space will benefit from Charles Pappas's revealing account of lesser
known characters, objects, and events from the Apollo era. Part memoir,
lunar program history, and layperson tour of spaceflight technologies,
Pappas draws readers in with his snappy prose and infectious excitement
about an era that has not gone by. The arc of human history was in fact
influenced by people who made the lunar landings possible and whose
positive imprint is evident in our world today.

Among the inventors and program managers who appear in his
fast-paced stories, Pappas shines a special light on the brilliant engineer
Robert Goddard. His liquid-fueled rockets sprung from a starry-eyed
imagination that helped realize his outrageous dreams (so said the *New
York Times*) of human spaceflight and turned Earth orbit into an arena
of workaday global communication, commerce, and data analysis. As
Pappas shows, fantastical visions often led to such prosaic—yet fun-
damentally transformative—new realities in the space age. Thus efforts
to design jet shoes, which cosmic pioneers would use to zoom towards
setting stars, yielded the Manned Maneuvering Units that astronauts
now use to operate outside of the International Space Station. Similarly,

Buck Rogers dreams of ray guns help inspire laser technology that may yet yield space-based weapons but in the meantime has changed the way commodities are tracked, metals fused, and eyes repaired.

Historians of the US space program know well that the demanding conditions of human spaceflight galvanized engineers to work quickly and invent things that became commonplace. So too does Charles Pappas, whose stories of spinoff technologies whirl with greater energy than standard historical accounts and prompt a satisfied mutter: "so that's the real story of the space pen, memory foam bed, and Tang—the sweet elixir of moonstruck Americans and now of people across Asia." These spinoffs, however, did not simply excite consumers. Many saved countless lives on Earth: emergency kits stock hypothermia-combating Mylar blankets, firefighters wear flame-resistant materials and breathe from high-tech air tanks, and sappers use solid-rocket fuel to neutralize the deadly global scourge of landmines one at a time.

This book entertainingly shows there was some truth to boosters' claim that human spaceflight would excite people's quest for knowledge. The emphatically judicious may discount the enduring urge to uncover UFOs, but they might enjoy Pappas's account of that quest and of the more conventional astronomical search for extraterrestrial life.

Whether you nostalgically commemorate or skeptically regard America's lunar landings, the half-century anniversary of Apollo 11 is an occasion for sober reflection. Or if you prefer recapturing the youthful excitement and adult wonder about human spaceflight, keep turning the pages and let Charles Pappas show you how that one giant leap changed our lives in ways big and small.

James Spiller
Professor of History
The College at Brockport, SUNY
Author *Frontiers for the American Century: Outer Space, Antarctica, and Cold War Nationalism*

Introduction

When I was about nine years old, I shook hands with a god. It was a lowercase "g" god, but you can't be picky about the deities you meet. My parents took me to the Capitol Building in Washington, DC, on a typical summer's day, when the air is a soaking hot washcloth tied over your face. This is where I found myself staring up into the face of Virgil "Gus" Grissom. Even at my single-digit age, I knew he had been the second American shot into suborbital flight on the Liberty Bell 7, and he had just come back a few months after his stint as the command pilot for Gemini 3, the first manned Project Gemini flight, which flew on March 23, 1965.

His three orbits and 80,000 miles around the Earth on the Gemini mission were overshadowed by another event inside the spacecraft. Grissom had smuggled a corned-beef sandwich on board. I didn't know that some of the very congressmen swarming around him right at that moment had held an angry hearing over that unauthorized snack. All I remember is looking up at that smooth, uncreased face with hair as black and flat as the top of an anvil. His hand formed a warm bear paw over mine. His nails, I noticed, weren't chomped down;

Virgil "Gus" Grissom, one of the original Mercury Seven astronauts, and the second American in space. NASA

Where the sidewalk ends: the reduced-gravity walking simulator. NASA

my own were anxious stubs. His smile was easy and loose, exactly what I had hoped for. He asked how I was doing, and I stammered something in the falsetto of preadolescence. Soaring on the flying-carpet high of awe, I was scared that letting go of the hand that had been in outer space and steered spaceships would turn the moment into dust.

Grissom was as heroic-large as anyone on Mount Olympus and Valhalla, although officially no member of the original Mercury Seven could be taller than 5 feet, 11 inches, and Grissom was the shortest of them all, at 5-foot-7. He was the embodiment of all the Tom Swift books I had devoured over and over, especially *Tom Swift and the Visitor from Planet X*, and *Tom Swift and the Asteroid Pirates.* The evidence of my fascination with them was the stains left from the grape popsicles and orange push-ups I ate while hunched over their pages. His presence and deeds in low orbit were a signet ring of space and adventure, and infinite possibility impressed on the soft wax of a child's mind.

Buzz Aldrin on the moon. NASA

The next few years brought maps of the solar system taped on all four corners to the yellow wall in my bedroom, my eyes always drawn to fat Saturn and its hula-hoop rings. That, and more Tom Swift books, *Star Trek*, and even on occasion, I am mortified to say, *Lost in Space*. It all culminated in 1969 with those words on the TV screen, LIVE FROM THE SURFACE OF THE MOON, and Walter Cronkite saying "Man on the moon!"

In my mind's eye, all those personal screen grabs—Grissom towering over me, the astronauts in their shiny suits boarding their capsules, the Saturn V knifing into the air with a blowtorch of thrust roaring out its back, Armstrong slowly walking on the moon as if treading through

President Nixon visits the Apollo 11 crew in quarantine. NASA

invisible water, Cronkite exulting on the air next to Wally Schirra, the Apollo 11 crew stuck in a Mobile Quarantine Facility peeking out through a rectangle of glass at President Nixon—are as iconically static as Mount Rushmore or *Washington Crossing the Delaware*. Motionless in their state of suspended animation, these images reflected a space race that was a 1-2-3, A-B-C, linear progression of milestones, each fitting on top of the one before as neatly as bricks in a wall.

The reality is, the space race was anything but well-ordered. Getting to the moon was a rumble of chaos, detours, inspiration, dead-ends, courage, failure, paranoia, dreams, whims, and brainstorms. The reality was less warp drives and phasers, and more diapers and crumbs.

Time-lapse images of a Gemini launch. NASA

One Giant Leap isn't an official history of the Apollo era, but rather snapshots of the less appreciated characters, objects, and events from that time: the unappreciated genius of Robert Goddard; the monstrous complexity of the Saturn V rocket; the classified plan to build a nuclear rocket; the clandestine scheme to explode a nuke on the moon; the legend of the space pen; the ingenious weapons put on board Russian spacecrafts; the mystery of space ice cream; and the spin-offs of NASA technology we live with every day and barely acknowledge.

Many of the people who defined that era in sharp, unforgettable strokes are gone now: Armstrong, Cronkite, Wernher von Braun, and of course, Grissom himself, killed in the Apollo 1 fire on the launchpad in 1967. He and the others are still as magic to me now as on that humid summer day in Washington, DC. Maybe more so, because now I understand they were heroes—not because they pursued power or acquired wealth, but because they gave the best part of themselves to something larger and more lasting.

CHAPTER ONE

Fly Me: Saturn V

THERE WAS NOTHING LIKE IT BEFORE AND THERE HAS BEEN NOTHING like it since. It is the equal of any painting hanging in the Louvre or any monument standing in Rome. It probably shouldn't have worked the first time it flew, but it did.

The Saturn V rocket was a multistage, liquid-fuel dispensable rocket used in the Apollo and Skylab programs. That one sentence, while perfectly accurate, tells no more about the missile than saying that Jason's mythical ship, the *Argo*, was made of oak.

Better to explain why this ship was magic: The *Argo* was designed by Athena, the Greek goddess of wisdom. Zeus gifted Jason with an enchanted chunk of speaking timber from Dodona, a site second only to the Oracle at Delphi in sacredness. Jason stuck it to the ship's prow, where it offered divine premonitions and prophecies.

Instead of the supernatural power of Athena and Zeus, the Saturn V had the industrial might of Boeing, McDonnell Douglas, and North American Aviation. Instead of Jason, the rocket had Wernher von Braun, the German-born engineering mastermind behind the V-2, the first ballistic missile that rained on London and other Allied cities. Under his efficient supervision, about 1,402 V-2s were launched at Britain, with more than 500 striking London. Von Braun administered the V-2 construction at the forced-labor Mittelwerk plant beneath Kohnstein Mountain, near the Dora concentration camp, where most of the 60,000 Mittelwerk workers were imprisoned. The missiles killed an estimated 2,754 civilians in London, with another 6,523 injured. Roughly 12,000 forced laborers and concentration camp prisoners died in the production of the weapons. We should never forget that the glory of the moon landing had some of its roots in immeasurable misery.

Formal development on the Saturn V began in January 1961, with von Braun overseeing its construction; Boeing, to assemble the Saturn V's first stage; North American, the second stage; and McDonnell Douglas, the last. Its design moved almost as fast as the Saturn V itself, going from paper pad to launchpad in six years, with its first uncrewed flight on November 9, 1967. It arrived in three separate stages at Cape Kennedy (the area was only "Cape Kennedy" from 1963 to 1973, before it reverted back to Cape Canaveral). One stage came by barge; the next, by a converted navy landing ship; and the third, aboard the *Super Guppy*, a puffed-up interpretation of the Boeing 377 Stratocruiser. All told, the massive rocket comprised three million parts furnished by twenty thousand different companies, universities, and government facilities. The manual for inspecting the Saturn V's components alone exceeded thirty thousand pages.

Von Braun wanted to test each part of the rocket individually, an understandable caution since the rocket, in size, power, and complexity, was unlike anything ever built before. Even today, after decades of accumulated expertise in rocket building, the sheer intricacy of these missiles almost guarantees that some must fail spectacularly, like the disastrous experience of SpaceX's Falcon 9 in June 2015. The $63 million rocket exploded on the launchpad, destroying the rocket and its cargo, a communications satellite owned by Israel-based Space Communication.

But with NASA's hurry-up schedule to plant humans on the moon before the end of the decade, George Mueller—NASA's director of the office of manned spaceflight—decided to forgo the methodical, step-by-step process von Braun wanted, and try out the entire rocket all at once.

Its specs today still elicit a frisson of awe: 363 feet tall; liftoff thrust of 7.5 million pounds; ferrying capability of 260,000 pounds to low Earth orbit; 960,000 gallons of propellant. The most powerful part of the Saturn V was the five F-1 rockets located underneath the first stage that, at liftoff, burned through a total of 500,000 pounds of fuel. With all three stages fully fueled, it contained 960,000 gallons of propellant, and

Wernher Von Braun in front of the Saturn V rocket that would be used for the Apollo 11 moon landing. NASA

The Saturn V for the Apollo 11 lunar mission makes the long, slow journey to the launchpad. NASA

weighed about 5.6 million pounds, around twenty-nine times the capacity of a Boeing 787-8 Dreamliner. (The two rocket boosters that pushed space shuttles into low Earth orbit, by contrast, put out just one-third as much thrust.) When the Saturn V booster launched, buildings shook and windows broke. The press building 3 miles away rattled like a children's toy, the sound there cresting at 120 decibels—just a bellow or two below 135 to 140 decibels, the standard threshold of pain for loud noises.

The Saturn V was so special that it even had its own house. Completed in 1966, the Vehicle Assembly Building (VAB) at the Kennedy Space Center was built just to shelter the Saturn V during final assembly before launch. Following its completion in the VAB, the rocket would be slowly towed to one of two launchpads, either 39-A or 39-B, about 4 miles away.

A Saturn V on the massive crawler-transporter near the Vehicle Assembly Building. NASA

One of the largest buildings on Earth by area, the VAB sits on 8 acres, spans 348,480 feet, and stands a staggering 525 feet tall by 518 feet wide. Into it was poured 65,000 cubic yards of concrete, and it was constructed with 98,590 tons of steel. To allow the rocket to actually leave the VAB, its doors, the largest in the world at 456 feet high, had to be fully open, a process that itself consumed forty-five minutes. On one side of the VAB is painted the largest American flag in existence, measuring a whopping a 209 feet tall and 110 feet wide. These are dimensions that quantify the confidence and hubris of the age.

To move the Saturn V was a feat of engineering almost as astounding as building it. Transferring it from the VAB to either of the designated launchpads, 39-A or 39-B, required the specially built crawler-transporter (CT). Custom-built by the Marion Power Shovel Co. of Marion, Ohio, the CT possessed the squat girth of a low mountain range. It measured 131 feet long by 114 feet wide, and weighed approximately 6.6 million pounds (equal to about fifteen Statues of Liberty). Its height varied, from 20 feet to 26 feet, depending on what was required. Its 16,375-horsepower engines can transport 18 million pounds, or twenty fully loaded 777 airplanes, moving along a track just under the stately if not speedy 1 mile per hour to the launchpad. The first mover, CT-1, has traveled around 1,960 miles in its life to and from the pads, while its sister unit, CT-2, has voyaged 2,207 miles.

Combined, the Saturn and crawler-transporter were so heavy (about 12 million pounds, give or take) that when they rolled out of the VAB to make the commute to the launchpad, they required a special road that could endure its crushing mass and not collapse. On May 20, 1969, when the stacked stages of the Saturn V rocket for the Apollo 11 mission left the VAB atop a crawler-transporter, they traveled along a short road topped with river rock from Alabama. The 18,159-foot-long "crawler-way" was designed by the US Army Corps of Engineers, and then built from November 1963 to August 1965, for about $7.5 million. The path's original river-rock form was rounded and smooth, more like ball bearings that would wiggle around rather than squish together, thereby building even more overwhelming pressure on the surface.

The crawlerway has four separate layers, including a 3.5-foot sub-base, a 3-foot crushed-aggregate base, a blacktop sealer, and finally, a top layer of rocks that runs about 8 inches thick on curves and 4 inches thick on the straightaway. Rock used for the road must rank an 8 on the Mohs scale of mineral hardness. (Diamond on the scale scores a hardness of 10.)

"When you launch in a rocket, you're not really flying that rocket," said astronaut Michael Anderson, who died on the ill-fated Columbia space shuttle in 2003. "You're just sort of hanging on."

Hanging on is probably an understatement when you're hurtling through the atmosphere sitting on top of five massive Rocketdyne F-1 engines cranking out thrust equivalent to the combined power of thirty Boeing 747 jumbo jets. The first stage of the Saturn V burned 20 tons of fuel per second. At that rate, in just a tenth of a second, it consumed

Taller than the Statue of Liberty: the Saturn V. NASA

ten times the amount Charles Lindbergh used for his entire trip crossing the Atlantic.

Everything about the Saturn V was oversized, including its sound. One of the loudest noises ever recorded, at 204 decibels (human eardrums can rupture at 150 decibels), the rocket's sound at launch by one estimate was roughly five hundred times as loud as a lion's maximum roar.

The also-ran, the Buffalo Bills of the oversize rocket race, was the Soviet Union's N-1 rocket. Designed to hurl cosmonauts to the moon during the space race, the enormous rocket almost measured up to the Saturn V, at nearly 345 feet tall. What it lacked in height it made up in brute power: Its first stage produced almost 30 percent more thrust than the Saturn V. At launch, it weighed 6.1 million pounds, and should have been able to lift payloads of up to 95 tons to space, allowing it to comfortably send cosmonauts to the moon. *Should have* are the operative words here, because the N-1 exploded at all of its four attempted launches between 1969 and 1972.

Probably intended to leach some of the glory away from the forthcoming moon landing, the July 1969 launch/explosion of an unmanned N-1 took place just days before Armstrong, Aldrin, and Collins took off. The blast of 1,496,000 pounds of liquid oxygen and kerosene was akin to about 1 kiloton of TNT (the blast at Hiroshima was equal to 15 kilotons of TNT). Some other sources say the blast was only equivalent to about 250 tons, but either way, people reported that their windows shattered as far as 25 miles away, making it one of largest non-nuclear, man-made—if accidental—explosions in history.

The Saturn V powered thirteen missions into space—the first dozen for the Apollo program, and the last, and thirteenth, hefting the McDonnell Douglas Skylab into orbit. All occurred without explosive incident. The most eloquent testimony to its durability came during the launch of Apollo 12 in late 1969. Thirty-six and a half seconds after liftoff, a lightning bolt hit the vehicle when it was about 1.5 miles up. Moments later, at an altitude of about 3.5 miles, another burst of lightning sizzled the ship. The power in the command module died. Thinking fast, an environmental control engineer on the ground recalled that there was a little-known switch in the command module that would shift the systems

to an auxiliary setting. The astronauts followed his inspired recommenda-
tion, the power immediately flowed back online, and the mission continued to its planned landing on the moon's Ocean of Storms.

A Saturn V hasn't flown since 1972. More years have elapsed since
its terminal voyage in 1972 and now than between Lindbergh's flight
across the Atlantic and the moon landing. In that rift of time, the first
GPS satellite, Navstar 1, was launched. Skylab, the first American space
station, crashed back to Earth. Pioneer 11 flew past Saturn. Voyager 1
photographed Jupiter's rings. Two space shuttles exploded. The Soviet
Union launched the Mir space station. And the United States began
building the International Space Station (ISS). A robot space probe
landed on an asteroid. The Mars Pathfinder's Sojourner became the first
operational rover on another planet. And yet for all their wonder, for all
their accomplishments, the luster of a Saturn V does not diminish in
their wake, but grows.

In 2014, the Lego Group solicited ideas from its fan base for new
Lego sets. Among the most popular submitted was one for the moon
rocket. In 2017, Lego released a 1:110 scale Saturn V. At 1 meter in
height, it was the tallest model the toymaker had ever produced, and its
1,969 pieces cleverly evoked the year humans first landed on the moon.
Revell Inc. once sold a 4-foot-tall version of the Saturn V rocket; a copy
of the model stands at the Smithsonian National Air and Space Museum
in Washington, DC.

Toys can bring joy, but the real thing ignites wonder and awe. Unused
but complete Saturn V rockets are on display at NASA's Johnson Space
Center, NASA's Kennedy Space Center, and the US Space & Rocket
Center. The sight of these metal giants in slumber raises goose bumps
like an ice cube run over the skin. The rocket is as true an emblem of its
time as the Wright Brothers Flyer (the first powered flight), the Boeing
B-29 Superfortress (the first to fly around the world nonstop), and the
Boeing 747 (the original jumbo jet) were of theirs. It represents the best
of engineers and the brightest of designers.

Yet for all of the millions of dollars and man-hours poured into its
development, the Saturn V rocket was always intended to be as disposable as a Kleenex. After pitching humans toward space, the rocket's first

Apollo 11 lunar module ascent stage, photographed from the command module.
NASA

stage detached from the rest and fell into the ocean. Then the second stage's five rocket engines ignited. (Shortly after, the emergency escape rocket on top of the vehicle, which could only be used below 19 miles altitude, was ditched.) At slightly more than nine minutes after launch, the second stage was cast off. Ten seconds later the third stage's rocket engine fired up to slide the crew into orbit around the Earth. After close to two and a half hours, the third stage rocket, the bottom end of the second stage, was jettisoned. Less than two hours after that, the third stage coasted away into space.

Much of the Saturn rockets lie quietly deep in the Atlantic, where time and salt slowly dissolve them, first from memory, and then, existence. Perhaps NASA always intended to leave those spent rocket stages there, a burial at sea befitting their stature.

But Jeff Bezos had other plans. In addition to founding Amazon. com, the Sears & Roebuck of the post-digital world, the billionaire also runs the spaceflight company, Blue Origin. Almost forty-four years after Apollo 11, he launched expeditions to salvage some of the stages in 2013. Even though Bezos found the likely location of various Saturn rocket remnants in a fifteen-minute-long Internet search, it took a crew of sixty three arduous weeks to find and haul up the remnants from their resting place 3 miles deep in the Atlantic. The fragments included parts of thrust chambers, gas generators, heat exchangers, fuel manifolds, and dozens of other artifacts. A scarred Apollo 12 F-1 engine was recovered, as well as the F-1 engine's thrust chambers from the Apollo 11 mission. Some of these pieces are on display in museums such as The Museum of Flight.

Nearly half a century since the last Saturn was launched, its descendants barely stack up.

Consider: At 229.6 feet, SpaceX's Falcon Heavy stands almost 140 feet shorter than the Saturn V. Its liftoff thrust of 5 million pounds is about two-thirds that of the Saturn. Blue Origin's New Glenn stands taller than the Falcon, but still almost four stories shorter than Saturn, and its 3.9 million pounds of liftoff thrust are close to half that of the Apollo 11 rocket. Despite the ballyhoo of more-nimble private businesses running circles around bureaucracy-dulled NASA, it's the government agency that stands to build the only true heir to Saturn V's legacy and status. NASA's Space Launch System (SLS) stands 322 to 365 feet tall, depending on the configuration needed. It's powered by up to 11.9 million pounds of thrust, or nearly 55 percent more than Saturn V could flex.

When I see the Saturn V's numbers, they give me no more sense of the rocket's power than saying "nine symphonies, thirty-two piano sonatas, one opera, and five piano concertos" gives me the sense of Beethoven's music. Standing next to the never-launched Saturn V at the US Space &

Rocket Center in Huntsville, Alabama, I recalled what Neil Armstrong said about looking back at the Earth from the moon: "I didn't feel like a giant. I felt very, very small."

Nothing modern can describe the Saturn V. The only fitting inscription for its legend is something written just a few years after the first practical steam locomotive was invented:

> And on the pedestal, these words appear:
> My name is Ozymandias, King of Kings;
> Look on my Works, ye Mighty, and despair!

CHAPTER TWO

Getting Off the Ground:
Robert Goddard

It was the most satisfying "You were right" in history.

Three days before Apollo 11 reached the moon and Neil Armstrong became the first human to tread across its gray surface of dead volcanoes and impact craters, the July 17, 1969, edition of the *New York Times* printed the mother of all corrections. "A Correction" was printed to a less-than-stellar editorial printed nearly fifty years before, in 1920.

Assessing Robert Goddard's seminal work, "A Method of Reaching Extreme Altitudes," an editorial under the heading "Topics of the Times" suggested—with a cherry of snark on top—that "professor Goddard with his 'chair' in Clark College and the countenancing of the Smithsonian Institution, does not know the relation of action to reaction; and of the need to have something better than a vacuum against which to react—to say that would be absurd."

This scoffing, along with the proto air quotes around his academic status of "chair," ended with a caustic flourish: "Of course, he only seems to lack the knowledge ladled out daily in high school." Almost a century later, you can still hear the slow clap and the mic drop.

The editorial forty-nine years later stated that "it is now definitely established that a rocket can function in a vacuum as well as in an atmosphere. The *Times* regrets the error." Goddard was the Nolan Bushnell of his time, founder of Atari and creator of the early video game, *Pong*. Its two-dimensional graphics were as rudimentary on a screen as tic-tac-toe is on paper, but *Pong* was nonetheless the provenance *Tetris*, *Minecraft*, and *Grand Theft Auto* all shared.

Or, instead of Bushnell, perhaps Nikola Tesla would be a fairer head-to-head comparison. The Serbian inventor sailed for America with four cents in his pocket and calculations for a flying machine in his head. He pioneered the first hydroelectric powerplant induction motor, and his work in alternating current laid the groundwork for modern wireless data. But for all his prophetic genius, Tesla was a secretive paranoid who claimed he had received signals in his Colorado laboratory from another planet, a claim that was met with derision in some scientific journals.

Goddard wrapped himself up in a cocoon of enigmatic privacy as dense as Tesla's.

Maybe his self-generated protective force field came about because he was often as alone on Earth as his rockets would someday be in the sky. As a child Goddard was so often sick that he was held back from school for undisclosed illnesses, to the point where he wasn't a sophomore in high school until he was nineteen, and didn't graduate—albeit, as valedictorian—until he was a ripe old twenty-two. Goddard earned a BS from Worcester Polytechnic Institute in Massachusetts, and his PhD in physics from nearby Clark University. (He later taught at the Worcester Polytechnic Institute of Clark University.)

In 1913, at age thirty-one, he became seriously ill with tuberculosis. When his prognosis suggested his life would be a transitory one, he focused on his rocket dreams and continued to work. And work. And work.

The moment after which there was no turning back for Goddard came about much like the story of Isaac Newton and the falling apple on the crisp afternoon of October 19, 1899. When he was seventeen, Goddard later reminisced, he climbed a tall cherry tree in his family yard to cut off dead limbs. Once he had scampered up near the top, he looked from his perch toward the east. "I imagined how wonderful it would be to make some device which had even the possibility of ascending to Mars," he wrote. "I was a different boy when I descended the tree from when I ascended, for existence at last seemed very purposive." Years later, the man who smoked cigars, played the piano, and painted for relaxation, still had the saw he'd taken up in the tree that day. For the remainder of his life, the boy who came down that tree celebrated October 19 as his own holiday. He called it "Anniversary Day."

Dr. Robert Goddard at Clark University. NASA

Goddard is credited with an astonishing total 214 patents, of which 131 were filed after he died. In 1914, though, Goddard received two key US patents, without which modern rocketry—like the Jupiter-C rocket that lifted the Explorer 1 satellite into orbit, or the Mercury-Redstone 3, that elevated Alan Shepard into suborbital flight, or the Saturn V with its 7.6 million pounds of thrust that shot mankind in the form of Michael Collins, Neil Armstrong, and Buzz Aldrin to the moon—is unthinkable.

Goddard's accomplishments have a Da Vincian gleam, of someone so far ahead of the curve he may as well have been developing a steam engine six thousand years ago on the grasslands of Ukraine and Kazakhstan, when horses were first being domesticated.

The first of Goddard's 1914 patents was for a rocket employing liquid fuel, and the second was for a two- or three-stage rocket using solid fuel. To convey how far ahead of his time he was in 1914, ponder this: The Wright Brothers had flown at Kitty Hawk, North Carolina, just eleven years before. In 1914, Willis Carrier patented an air conditioner in the United States. The inaugural nonstop solo flight across the Atlantic by Charles Lindbergh wouldn't take place for thirteen more years.

When Goddard fired his first liquid-fueled rocket on March 16, 1926, he did it on the open expanse of his aunt Effie's Auburn, Massachusetts, farm (now Pakachoag Golf Course). It burned for about twenty seconds before taking off, melting part of the nozzle.

According to his wife Esther, Goddard took in the boiling tableau and exclaimed, "I think I'll get the hell out of here!" Nell—Goddard's nickname for his rocket—rose 41 feet, and flew 184 feet at an average speed of 60 miles per hour. "It looked almost magical as it rose, without any appreciably greater noise or flame, as if it said 'I've been here long enough; I think I'll be going somewhere else, if you don't mind,'" Goddard later wrote in his journal.

Despite the snippy *New York Times* takedown, Goddard flared with the brilliance of a Roman candle. On March 16, 1926, he launched the first liquid-fuel-powered rocket. That was the opening of an Edison-like production line of marvels. He flung the first scientific payload (a barometer and a camera) into the sky in in 1929; he developed an internal guidance system (a gyro apparatus) for rockets in 1932; he fired off a liquid-fuel rocket that journeyed faster than the speed of sound in 1935. Even the invention he never pursued still possessed a Tomorrowland gleam. His patents—#2,488,287, which is called "Apparatus for Vacuum Tube Transportation," and #2,511,979, which is labeled "Vacuum Tube Transportation Systems"—imagined a kind of magnetic-levitation vehicle inside a vacuum tube that could move as fast as whatever maximum speed the human body could endure.

And yet Goddard labored on in a blind spot of science and culture, one chained to the era's limited understanding of what rocketry might bring, and partly due to his self-imposed isolation that crippled him more than any of the critics taunting him with names like the "moon professor." After the Smithsonian published his "A Method of Reaching Extreme Altitudes" (issued in Smithsonian Miscellaneous Collections, Volume 71, Number 2, 1919), the media picked up on its implications of rockets reaching the moon. Although Goddard focused solely on

Dr. Robert Goddard and "Nell," the first liquid oxygen–gasoline rocket in 1926. NASA

solid-fuel rockets in the article, he also offhandedly described a highly theoretical multistage, unmanned, liquid-fueled rocket that might be capable of reaching our closest interplanetary neighbor. That story traveled faster and farther than his rockets ever could, with the *Times* famously mocking him, and enthusiasts clamoring to join the crew of his impending lunar journey. Goddard even received interest from Hollywood, where an agent telegrammed him:

> *Would be grateful for opportunity to send message to moon from Mary Pickford on your torpedo rocket when it starts.*

Publicity, whether in the form of an unwelcome editorial or an unwanted fame, curled him inward even more, a leaf raked by the cold of autumn. Goddard would be a think tank of one for all of his life. He forced his handyman assistants to sign oaths that promised they would never expose the details of the work they did for him. To do so would result in their abrupt termination. He revealed the 1926 launch only to those who had a kind of need-to-know status, including Charles Abbot, the director of the Smithsonian Astrophysical Observatory. He likely revealed it only because the institution had supported him with a $5,000 grant in 1917, and had legitimized him with the 1919 publication. He didn't go public with the 1926 launch until his Smithsonian-published paper, "Liquid-Propellant Rocket Development," a decade later, in 1936.

Sickness and scorn and celebrity joined with his competitiveness to induce a kind of functioning, low-grade paranoia. With the rise of groups such as the American Interplanetary Society and the German Society for Spaceship Travel (*Verein fur Raumschiffahrt*, of which Wernher von Braun, the brainiac behind the Saturn V rocket, was a member), Goddard feared his forward-thinking innovations would invite thievery as much as mockery. Like a prophet in Bible times, he repaired to the wilderness. In this case, that meant Roswell, New Mexico, and its surrounding desert, to build and launch rockets. There he would attempt forty-eight flights, with thirty-one of them lifting off, including L-13, which peaked at an elevation of 1.7 miles, the highest any of his rockets flew. At one point Goddard packed a gleaming aluminum and stainless-steel rocket, slightly

Goddard with his double-acting-engine rocket, in 1925. NASA

more than 15 feet long, into a wooden crate and shipped it to the Smithsonian for posterity. Then he went on to demand that his rocket never be shown in any manner without his permission. It was moved deep into the bowels of the rust-red-sandstone Smithsonian Castle and left there, unseen and unappreciated for more than a decade.

Goddard divulged to Charles Lindbergh—who had a keen interest in abandoning propeller-powered planes for jet-driven ones—that a stipend of $25,000 a year for four years would allow him to accomplish in that short span what would otherwise take him a lifetime. Lindbergh took the hint, and consequently brokered a meeting in Wilmington, Delaware, with industrialist Henry du Pont and a select group of engineers. But once the engineers began to scribble notes when Goddard revealed details about his rockets, he shut down, and all the light Goddard had begun to share went out of the conversation. After the meeting, Lindbergh flew Goddard back from Wilmington to New York, the first time Goddard had ever truly flown—in body, if not imagination. After soaring as high as his rockets and then diving so low the plane almost scraped treetops, Goddard wrote that the experience was a "a good test of my nerves."

Lindbergh later approached his friend, Harry Guggenheim, son of philanthropist Daniel Guggenheim. Soon the Daniel and Florence Guggenheim Foundation offered Goddard an initial grant of $50,000, followed over the years by $138,500 more.

Even with friends like Lindbergh and benefactors such as the Guggenheims, Goddard kept his own counsel, often to his supporters' irritation. In a letter Lindbergh penned to Harry Guggenheim in 1936, the aviator groused about Goddard's more abstract pursuits: "I would much prefer to have Goddard interested in real scientific development than to have him primarily interested in more spectacular achievements which are of less real value." (Likely Goddard's focus on pure rocketry rather than jet-powered airplanes had frustrated Lindbergh.) Five years later, as the United States was about to enter the war that would accelerate every aspect of manned and unmanned flight, Brig. Gen. George H. Brett, chief of Materiel Division, US Army Air Corps, wrote a letter to Goddard rebuffing his rocket research proposals:

R. H. GODDARD.
ROCKET APPARATUS.
APPLICATION FILED OCT. 1, 1913.

1,102,653.

Patented July 7, 1914.

The 1914 patent for Goddard's rocket apparatus. USPTO

The proposals as outlined in your letter . . . have been carefully reviewed . . . While the Air Corps is deeply interested in the research work being carried out by your organization . . . it does not, at this time, feel justified in obligating further funds for basic jet propulsion research and experimentation.

When the war started, Goddard and Esther left Roswell in 1942 to work with the US Navy and Curtiss-Wright Corp. as a consulting engineer. Just three years later, in August, 1945, he was dead of throat cancer, never having seen the heights that those who understood his work would reach.

His legacy was left in the hands of Esther, his de facto publicist and photographer throughout his life. She kept his memory aloft, filing 131 patent claims after his death, and assisting in publishing some of his writings. In 1951 she teamed up with the Guggenheim Foundation to file a joint claim against the US government for infringing upon Goddard's patents. Government moves slowly, and justice even slower, but in June of 1960, Esther and the Guggenheim won a $1 million settlement. It was another record, if posthumous, for Goddard: the largest patent settlement the government had ever made at the time.

The deserved tributes slowly poured in. On May 1, 1959, NASA formally established what would be the Goddard Space Flight Center in Greenbelt, Maryland, the organization's first spaceflight center. (Esther attended the dedication of the center on March 16, 1961, exactly thirty-five years after Goddard launched Nell from his aunt Effie's farm.) A few months afterward, the 86th Congress authorized the issuance of a gold medal honoring Goddard.

For all the accolades, for all the records, for all the benefactors, from the Smithsonian to Lindbergh, I think the *New York Times* editorial and all the other jibes had always been a stone in Goddard's shoe. Yet his response to it possessed a kind of wounded grace that defined his legacy. Shortly after the editorial, Goddard issued a statement to the Associated Press that stands as the last word. "Every vision is a joke," he said to one reporter, "until the first man accomplishes it. Once realized, it becomes commonplace."

Chapter Three

Timing Is Everything: Watches

EVER SINCE HUMANKIND BEGAN CHARTING THE PASSAGE OF TIME WITH shadow clocks and sundials, time itself has been compared to a dress-maker, a counselor, a stream, a fire, a bank, a prison, a thief, and a "reef upon which all our frail mystic ships are wrecked." What time is not, however, is a slave. It can only be measured, not mastered. Scrutinized, not subjugated. Not even by astronauts.

It's likely the first watch to make it into space was Russian. When Yuri Gagarin blasted off from the Baikonur Cosmodrome in Kazakhstan on April 12, 1961, he probably wore his personal timepiece: a mechanical, manual-wind Sturmanskije chronograph, which he would have received when he graduated from the Russian air force flight school. (Some argue he would have been sporting an upgraded version.) Sturmanskije, which translates as "navigator," was a brand manufactured first in 1949, unavail-able to ordinary comrades at the time.

While Alan Shepard became the first American in space, John Glenn became the first American to orbit the Earth—and, as far as is known, the first to wear a watch. Glenn, the US Marine Corps aviator who later became the oldest person to fly into space, at age seventy-seven, strapped a Heuer 2915A stopwatch over his space suit, with improvised elastic bands holding it in place. Never late in pouncing on the marketing opportunity, the Swiss luxury company claimed this made Heuer the first Swiss watchmaker in space.

Following Glenn, his fellow Project Mercury astronaut, Scott Car-penter, orbited the Earth three times in May of 1962 on the Aurora 7 mission, with a Breitling Navitimer pilot's watch. Before his flight, Carpenter (who had worn a Breitling Navitimer during the Korean War, and later, as a test pilot) asked the company to modify the watch by removing the tachymeter scale and adding a twenty-four-hour dial, in case he became stranded in space and needed to differentiate day from night. The modified watch performed well enough in space, but it succumbed to seawater during splashdown in the Atlantic.

The quest for a watch hardy enough to endure the journey from blastoff to orbit to splashdown continued. A few months later, Wally Schirra and Gordon Cooper strolled into a jewelry store in Houston to buy watches on their own dime. Choosing the Omega Speedmaster CK2998 to wear on their respective flights, they selected the manualwinding timepieces likely because of their branding as rugged, he-man chronometers for sports and racing, just the sort one would expect someone with the right stuff to wear. Then on October 3, 1962, Schirra, the only astronaut to fly Mercury, Gemini, and Apollo missions, achieved another kind of record not as often bronzed for posterity in the history books: the first to wear a Speedmaster in space.

In September, 1964, with the Mercury program drawing to a close after six total flights that shot six astronauts in space, the astronauts asked operations director Deke Slayton to choose an official, flight-approved watch that could be worn during training and, eventually, the missions themselves.

Slayton considered the request, then banged out an internal memo asking for a "highly durable and accurate chronograph to be used by Gemini and Apollo flight crews." A few days later, James Ragan, the engineer responsible for testing equipment for the US Navy's Sealab program, took up the chronological challenge with a relish. He mailed out a "request for quotations" to ten different watch brands, asking for two watches apiece. Just four answered NASA's appeal: Rolex SA, Omega SA, Hamilton Watch Co., and Compagnie des Montres Longines. Hamilton, for reasons left obscure in the historical record, provided a pocket watch. With just a fraction of brands approached volunteering to take part, Ragan upped his original request, asking the companies to send three watches each.

John Glenn in full Mercury suit and watch. NASA

It's little wonder NASA would need a dozen timepieces, as the casualty rate in the testing phases was always going to be high. Most watchmakers who didn't RSVP for the test-to-destruction regimen likely wanted to avoid the negative publicity that would leave their wares looking like also-rans. The nearly dozen tests included a forty-eight-hour stint at a temperature of 160 degrees F, followed by thirty minutes at +200 Fahrenheit; four hours at a temperature of 0 degrees Fahrenheit; ten days at temperatures fluctuating between +68 degrees Fahrenheit and +160 degrees Fahrenheit, in a relative humidity of at least 95 percent; and multiple shocks of 40 Gs, each eleven milliseconds in duration, with the watches moving in six different directions. Even more tests involving high pressure, acoustic noise, decompression, and vibrations were used to stress the timepieces to their breaking points.

The tests ended on March 1, 1965. The last man standing—or, more accurately, the last watch ticking—was the Omega Speedmaster reference ST105.003. The end result of this excruciating examination was NASA declaring Speedmaster "Flight Qualified for all Manned Space Missions" on the March 1, 1965. Only the makers of Tang might not have been impressed by the product-placement possibilities.

Since then, the iconic Omega has appeared on every manned Gemini and Apollo mission, as well as trips to Skylab and the Apollo-Soyuz test project. But other timepieces, taking advantage of chance and circumstance, made their mark. During the 1971 Apollo 15 mission, the crystal came off of mission commander Dave Scott's Speedmaster. Prepared for any contingency, Scott pulled out a backup, his own personal Bulova Chronograph, that he then used during his lunar walk. That last-second substitution earned Bulova the distinction of being the only privately owned watch yet worn on the moon.

The Russians, while not as commercially conscious as the Americans, nonetheless used anachronistic technology, which can be traced back to the portable clocks of the eighteenth century. The first watch to be worn in open space was the white-faced Strela (Russian for "Arrow") on the wrist of thirty-year-old cosmonaut Alexey Leonov, when he left the Voskhod 2 spacecraft on March 18, 1965, for the first spacewalk (or EVA, extravehicular activity, as it is technically known). This particular brand had first been issued to Soviet pilots in the 1950s and became the standard cosmonaut timepiece until it was withdrawn in 1979. Today, Strela has been reborn as a German brand in Munich.

Watches held a special place in astronauts' hearts. "A gentleman's choice of timepiece says as much about him as does his Savile Row suit," said Ian Fleming, the creator of James Bond (who has worn a Rolex, Seiko, and, most recently, an Omega watch in the movies). After completing a mission, astronauts were supposed to hand their watches back, like they would any other piece of equipment. Over time, though, they began to look at the watches as extensions of themselves, like wedding bands or class rings. The problem was so pronounced that by the time of Skylab, Deke Slayton, who had served as director of flight crew operations, threatened to cut off astronauts' time in the T-38 trainer if they didn't hand their watches back. Since time in the T-38 jet trainer was mandatory to qualify for spaceflight, Slayton was in effect promising to ground them. The watches found their way back into official hands.

Even so, the line between what belonged to NASA and what belonged to the astronauts was nebulous. Photos of Ed White's 1965 spacewalk show him sporting two watches of indeterminate origin. Ed Mitchell, some say, had no fewer than three watches on during his Apollo 14 flight. Eugene Cernan and Ronald Evans, flying on Apollo 17 the last and final Apollo mission, may have worn their own personal Omegas beneath their space suits, with the NASA-issued ones strapped to the outside. Whether they carried the watches into space to give as gifts or keep as an off-world memento, no one can be sure. Moreover, the personal watches weren't a departure from the norm; astronauts have smuggled everything on board from Scotch to lingerie.

The issue of who owned what came to a boil in 2012, when a checklist used by Apollo 13 commander James Lovell sold at auction for a record-setting $388,375. First there was the misdirected outrage over astronauts selling for personal gain historic memorabilia that rightly belonged to the government. In the calm that settled in after the moralizing storm, when the country realized no real harm was done, President Obama signed legislation that gave the Mercury, Gemini, and Apollo astronauts legal title to whatever items or equipment that was "not expressly required to be returned" to NASA at mission's end. The gear affected by the law covered myriad items, from toothbrushes to cameras to checklists to, of course, watches.

Those timepieces still have a luster as glossy as moonlight. Aficionados of astronaut watches view them with the hunger and fascination medieval traders did religious relics. Instead of coveting bones, skin, fingernails, and even heads of saints, these enthusiasts are driven to acquire the watches, even at wallet-rupturing prices, as long as they've rubbed against astronaut skin—like the elegant 18 karat gold Vacheron Constantin Ed Mitchell received as a gift after his return from space in 1971. That watch was auctioned at Christie's New York in 2016 for $62,500.

Anything worth wanting is worth stealing, especially a ticking object of desire. After the Aurora 7 flight, NASA sent Carpenter's water-damaged watch back to Breitling for repairs, but with its space-enhanced value now as high as Carpenter's flight, it mysteriously went missing and has never been found, despite Breitling's efforts to locate it. Buzz Aldrin's watch disappeared en route to the National Air and Space Museum (to which NASA had donated some fifty of the astronauts' watches in the 1970s). At least seven of the Museum's watches have been stolen in the years since, while they were loaned out to other institutions.

These enthusiasts are the reason why the makers of astronauts' watches keep reincarnating them years after their missions. Decades-old fame didn't stop Breitling from issuing a limited-edition version of the Navitimer in 2012 based on Carpenter's adjustments. Evoking the year of Carpenter's flight, the watchmaker manufactured just 1,962 of the watches. One was recently available on eBay for $5,499.

Similarly, TAG Heuer S.A. issued the Carrera 1887 SpaceX in 2012 to commemorate the fiftieth anniversary of John Glenn's flight in Friendship 7. It features an image of the Friendship 7 capsule on both the dial and across the display caseback. Not one to miss out, in 2016, Bulova made a faithful re-creation of Dave Scott's watch, powered by the brand's quartz chronograph movement.

The space race did more than just outfit a select few with upscale watches. It also changed how we keep time. This is because what was precise enough for astronauts themselves was not accurate enough for the often-invisible equipment that ultimately kept them alive.

Starting in 1961 at its Goldstone tracking station, NASA began using quartz oscillators to vastly improve the accuracy of velocity measurements, from 10s of meters per second to roughly 50 millimeters per second. Later, General Time Corp. developed a quartz crystal oscillator for the central timing equipment that formed a base for all Apollo mission timing functions. By vibrating up to 4,194,304 times a second, the quartz crystals could allow clocks to be accurate within a minute per year.

The technology soon trickled down to consumer clocks and watches, the electrically stimulated quartz crystals ensuring that millions would never be late again for anything from first dates to job interviews. What wood did for shelter and clay did for food, quartz did for time.

This democratization of time made the cheapest of watches as accurate as the best—a Timex is as correct as a Rolex. Now the lux brands, like Omega, fell back, hiring glamourous spokespersons—such as supermodel Cindy Crawford and film star George Clooney—placing products in popular entertainment (e.g., James Bond), and touting their long history and association with one-of-a-kind events, especially the space race.

We love the idea that an older, perhaps obsolete, tool can come to the rescue when the most advanced technology fails. Roman aqueducts will endure another millennium while steel bridges fall apart. Vinyl can deliver a far richer sound than digital music can never reach. Landline phones will work when their fussy successor, the cell phone, can't pick up a signal. A paper document can be kept truly secret while computer files are as open as a public park. Of course, this only ever happens in the movies.

Usually.

After the Apollo 13 oxygen tank blew up, severely damaging the craft and aborting the mission's moon landing, the crew limped back to Earth. Inside the lunar module (which had been designed to keep two of the astronauts alive for two days on the moon's surface, not sustain three people for four days on a trip back to Earth), the crew of Jim Lovell, Jack Swigert, and Fred Haise waited out their final hours. To keep the ship's dwindling power running as long as possible, they shut down nearly all power, including heat and the cabin clock.

When the ship neared Earth, Mission Control in Houston informed the astronauts they were coming in off course, roughly by 60 to 80 nautical miles. At that angle, once they hit the atmosphere, the ship would be swatted back into space. They would never return home.

There was one way back: manually adjust the Apollo 13's course with a precisely timed fourteen-second burn of fuel. A second too long or a second too short meant catastrophe.

Without normal navigation equipment available, the crew visually lined the ship up against the Earth's terminator, the line that cuts between daytime and darkness, and started the engine. Using his Omega Speedmaster, Swigert timed the burn of the rockets . . . 1 . . . 2 . . . 3 . . . At exactly fourteen seconds, the astronauts cut the engines off. The burn and its timing were perfect. For all their exaggerated utility, implied machismo, and eventual obsolescence, the old-time watches helped bring Apollo 13 safely home.

Chapter Four

Writing High:
Space Pens

IN THEIR SUBVERSIVE ARCHITECTURE OF SETUPS AND PUNCH LINES, jokes have a way of illuminating what stories, essays, op-eds, polls, tweets, retweets, call-in radio shows, and late-night talk-show monologues can't about a culture.

Take for example, the legend of the space pen.

When NASA started shooting men into space in the 1960s, it soon realized that ballpoint pens functioned poorly in conditions of zero gravity. Bringing its huge brainpower to bear on the problem, NASA directed its scientists to labor for a decade on the problem, spending $12 billion—an amount roughly equal to $100 billion today, or more than seven times the cost of the new $13 billion nuclear-powered USS *Gerald R. Ford* aircraft carrier. After dozens of false starts and thousands of man-hours of research, scientists ultimately developed a high-tech pen that could write smoothly and in all adverse conditions, even if the person holding it was inconveniently upside down, stuck in zero gravity, and exposed to temperatures ranging from below freezing to over 600 degrees Fahrenheit (nearly hot enough to melt lead). The stylus could also write on any surface, from coarse metal to smooth glass.

Meanwhile, the Russians, competing with—and usually besting—the United States on every level of the furious space race, had developed a writing device that was capable of all that and more. Even better, this superior implement cost the equivalent of mere pennies. The technically advanced implement that testified so effectively to the innate superiority of the Soviet system? It was called a "pencil."

For nearly fifty years or more, some version of this story has floated around with the ubiquity of pollen, its Aesop-like fiction capturing the creative and sometimes reckless spending of the United States compared to the brutalist, not to mention frugal, common sense of the Soviet Union.

Finding a permanent home on the Internet, including its own dedicated entry on the debunking site, Snopes.com (aka, the Urban Legends Reference Pages), and even warranting a mention on a 2002 episode of *The West Wing*, the space-pen parable was worth its weight in gold in BS. It was as much an urban legend as leaving a tooth in a cup of Coca-Cola overnight would dissolve it by the next sunrise. Who wouldn't believe a story that played on prejudicial tropes, like overeducated eggheads who reinvent the wheel, and government pencil pushers who, as another myth would have it, happily and irresponsibly shelled out $600 for a $15 hammer?

Like all legends, though, there is a nugget of truth in a moon full of misconception. In its early days, the Apollo program was deeply enmeshed in scandal that drove the space-pen brouhaha.

Originally American astronauts, like the cosmonauts, wrote with pencils, according to NASA historians. Ballpoint pens work poorly in space, because they rely somewhat on gravity to force ink out of the reservoirs located in their cartridges. Without gravity to give it a shove, there's nothing to push the ink from the cartridge outward to the ball. Which explains why the run-of-the-mill ballpoint pens don't write reliably upside down or on vertical surfaces.

That's also why in 1965, for the Gemini flights, NASA bought up thirty-four mechanical pencils from Tycam Engineering Manufacturing in Houston for the princely sum of $4,382.50 in total.

When the public found out that astronauts were using pencils priced at $128.89 apiece, there was the proverbial hell to pay. (The bar for what constituted a scandal then was low enough to limbo under.) Given that the average cost of a new car ran around $2,650 (a V-8-powered Ford Mustang coupe was $2,734), it was understandable when the public fumed, the newspapers inveighed, and Congress, which can resist attention-seeking behavior to the same degree ants can resist a picnic, got in on the act. Congressman John Wydler of New York, a member of the House Committee on Science and Astronautics, wrote a stern letter to NASA administrator

James Webb requesting a thorough investigation of the exorbitant pencils and their shocking price tag.

NASA proved its damage-control skills were probably better applied to potentially genuine tragedies like Apollo 13 than a kerfuffle over over-priced writing utensils. At first, agency representatives tried to explain that the mechanical pencils' initial cost was a reasonable $1.75 each. The additional charges only came into play because it NASA had to have an extra-strong outer casing of special fabric around the pencil so they could be attached to the inside of a spaceship, and more easily handled by astronauts wearing the thick clunky gloves covering their hands.

Despite the pandemonium over the mechanical pencils, the real Pencilgate slid by everyone in the public, media, and Congress. After Gus Grissom and John Young returned in Gemini 3 (impishly nicknamed "Molly Brown" in deference to Grissom's almost drowning on his Liberty Bell 7 flight three years before) in 1964, a government investigator, E. E. Christensen, wrote a letter concerning the astronauts' behavior to George Mueller, the head of the Office of Manned Spaceflight. In it, the investigator lists all the many items Grissom and Young smuggled on board, presumably for a financial windfall or a sentimental souvenir. Among the contraband the astronauts snuck onto Gemini 3 were an American flag, a diamond ring, Florentine crosses, a sandwich, and a brassiere. While none seems to have merited a firing squad or even a pink slip (the sandwich, however, did pose a threat of sorts), the one that caused the most consternation was pencils. As the letter put it:

> *Four Pentel pencils were taken aboard the flight. These are Japanese pencils with a nylon point. These were flight qualified. The problem, of course, is the 49¢ total cost. Deke Slayton, to whom I talked, was instructed to take very precaution in preventing this item from becoming public.*

No other item—including the wayward bra—warranted the letter's focus on keeping it out of public sight. The reason was simple: In the aftermath of the Tycam mechanical pencil disgrace, if all along NASA could have used pencils that cost about 12¢ apiece instead of $128.89—

made by our recent enemies and current competitors, no less—that would have toppled careers and gutted reputations.

Extortionately priced or not, pencils were in truth a lousy choice of writing tool. Tips could snap off, shooting debris into electrical instruments or human eyes, much as NASA feared crumbs from cookies and bread would. Sharpening them in zero gravity would just send shavings floating into the nooks and crannies of the electronics, where, because graphite conducts electricity, they might short-circuit vital instruments. On top of those drawbacks, pencils burn, which is a less-than-positive trait in space, where the enclosed environments, often with high oxygen content, are like a dream date between kindling and matches.

That's when Paul Fisher became a part of this immortal and evergreen story. Fisher on his own, and without NASA funding, set out to create the perfect writing tool for astronauts that would transcend the fearsome complications of gravity and shavings.

Following a stint working in an airplane propeller factory during World War II, Fisher labored in a pen factory, which inspired him to open up one of his own. Intrigued by NASA's dilemma, he sank $1 million of his own money into the problem. In 1966, the Fisher Pen Co. debuted its patented AG7, touted as the first "Anti-Gravity" pen.

AG7 was the first ballpoint pen designed expressly to operate in the antagonistic environs of space, where rogue pencil shavings, not the Borg or Cylons, could mean death and destruction. Employing a pressurized ink cartridge, the AG7 functioned underwater in zero gravity, immersed in other fluids, and in temperatures fluctuating from -30 degrees Fahrenheit to +250 degrees Fahrenheit.

Its ballpoint was made from hard-as-nails tungsten carbide. Inside, the ink was thixotropic, which meant when the pen was idle, the ink thickened, which prevented it from leaking. As soon as the pen started in motion again, however, the ink turned back into a smooth-running liquid. In other words, the Anti-Gravity pen worked like a ketchup bottle, its insides emerging after a few good shakes.

Fisher offered the pens to NASA, but after being burned by the negative coverage over the $128.89 mechanical pencils, the agency was understandably reluctant to touch them.

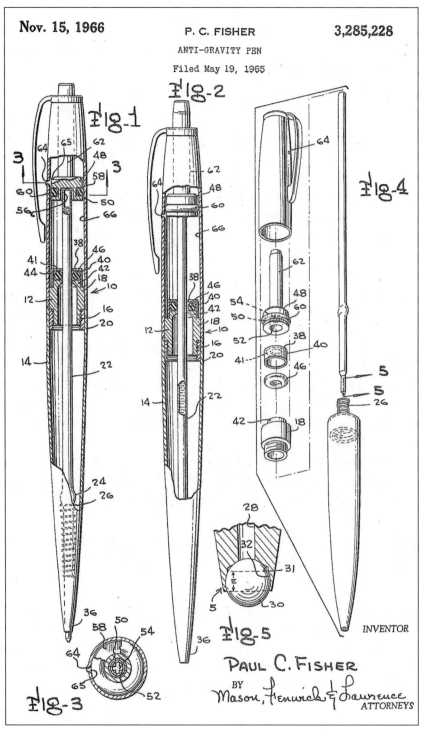

Fig-1

Fig-2

Fig-4

Fig-3

Fig-5

INVENTOR

Paul C. Fisher

BY

Mason, Fenwick & Lawrence
ATTORNEYS

Paul Fisher's space-pen patent. NASA

Happily, wiser heads prevailed. NASA accepted samples of the pens and subjected them to a gamut of trials equal in rigor to those suffered by the wristwatches tested. In the end, the agency bought approximately four hundred of the AG7 pens. The AG7 Fisher Space Pen was used on the Apollo 7 mission after eighteen months of rigorous testing by NASA. Because it bought the pens in bulk, NASA received a 40 percent discount, paying just $2.39 apiece instead of the $3.98 retail price. Like the way it de-branded Tang, NASA, in its "Handbook of Operational Equipment for Space Flight," officially referred to the AG7, with the inspired dryness of an end-user agreement, as a "Data Recording Pen."

Recognizing a good thing, the Soviets acquired one hundred of the pens, along with one thousand ink cartridges, in 1969, for its Soyuz space-flights. It received the same hefty price break as its capitalist competitor.

Why, if the original Soviet version of a writing tool was so much simpler, cheaper, and presumably safer, did the Russians fork over money for over-engineered frippery? The Russian space program used grease pencils, similar to crayons in that writing with them results more in smears and splotches than legible lettering. Instead of sharpening them to get more lead to write with, as you would a pencil, you just peel off another layer of paper on the pencil. This results in bits of paper floating around in low-gravity conditions, which are no safer on a Soviet ship than an American craft.

First used on the Apollo 7 mission that lasted almost eleven days, the space pen has written its way into history and folklore almost as widespread and enduring as the urban-legend counterpart. Fisher died in 2006, but the early promotional blurbs for his space pens, which promised it had a shelf life of one hundred years, may turn out to be truer than any copywriter envisioned. The AG7 was even the subject of an entire *Seinfeld* episode, in 1991's "The Pen."

When it comes to learning the ways of capitalism, the Russians proved to be as fast as a card cheat. In early 1998, two cosmonauts broadcast live via satellite from the Mir space station on the cable shopping channel QVC. Commander Anatoly Solovev and flight engineer Pavel Vinogradov hawked the $32.72 "Zero Gravity Fisher Space Pen" to show how it worked, 200 miles high above the Earth on Mir, offering it up as the first product "sold from space." Besides the pens, shoppers at home could purchase meteorites that were said to come from Mars, and three Russian-made Sokol KV-2 space suits, priced to sell at $25,000 apiece. The sale wasn't to line the cosmonauts' pockets; proceeds would instead go to the financially strapped Russian space agency, according to QVC.

It turned that there was more than one fable the space pen was connected to, a piece of now-forgotten space folklore that was even more dramatic. When Neil Armstrong and Buzz Aldrin were preparing the lunar module, the Eagle, to leave the moon, Aldrin saw that the ascent engine arming breaker push/pull switch had been busted. The likely culprit was one of the astronauts themselves, whose bulky space suit "backpack" had probably bumped into the panel and snapped off that particular switch.

It was much more than just an extraneous snafu that could be ignored. Mission Control confirmed that the open switch meant the engine was currently unarmed. If the engine couldn't be armed, Aldrin and Armstrong would be stranded on the moon to die slowly as their oxygen and life support dropped to zero. Thinking quickly, Aldrin whipped out his space pen and wrenched the switch open, successfully arming the engine and allowing the Eagle to take off. The mission and their lives were saved, thanks to the handy, do-it-all, write-anywhere pen.

It's a good story. It's such a good story that it almost hurts to read that years later, Aldrin confirmed he actually used a black felt-tip marker, made by the Duro Pen Company, to rearm the engine. But given the rich urban legend that revolved around the pen, it's wise to invoke guidance from the *The Man Who Shot Liberty Valence*: "When the legend becomes fact, print the legend."

CHAPTER FIVE

Stepping Out:
Jet Shoes

From Cinderella's enchanted glass slipper to *The Absent-Minded Professor*'s Flubber-augmented footwear, shoes are magic. "Shoes are paradoxical objects in that they constrict feet and yet free them to cover greater distances in space," wrote Isabel Cardigos in her book, *The Wearing and Shedding of Enchanted Shoes*. Need more proof? Witness Mercury and Hermes' winged shoes that needed a speedometer more than laces. Observe Dorothy's shedding her farm footwear for ruby red slippers that took her through Oz, both to the Wizard, and, ultimately, back to the Dust Bowl she called home. Remember Elvis warning others off his blue suede kicks with a sneer and a gyroscopic twist of the hips.

What was good enough for gods, princesses, and rock-and-roll stars was good enough for space explorers. Because buried in the NASA archives was John D. Bird's 1965 patent for jet-powered footwear that would have whisked its owners to Sputnik levels of altitude. Its initial description, though, was no more animated than a fruit bowl: "An apparatus for attachment to the feet of a person desiring extravehicular space locomotion having fluid thruster controlled by the toes of the person. Toe end heel attachment elements are utilized for securing a base that supports the thruster and a control circuit therefor [*sic*] to each foot. The control circuit is either electric, having a switch for energizing a solenoid valve, permitting fluid to flow through a nozzle, or fluidic, having a syring [*sic*] connected to a relay that operates a valve permitting fluid to flow through the nozzle."

The TL;DR version: it's Iron Man . . . IN SPACE.

Bird's patent ultimately changed the course of how we boldly went into the final frontier. Moonwalk, if you will, back to the mid-1960s. The space race was less than a decade old, after the Soviets had launched Sputnik 1 on top of a modified R-7 two-stage ICBM in 1957. Both countries had since accentuated the error part of "trial and error" in their efforts to successfully put living cargo into orbit, with the American Vanguard T exploding at launch in 1957, and the Russian R-16 ICBM igniting on the launchpad in 1960.

Yet despite the odds that any given rocket launch might become an impromptu fireworks display, on March 18, 1965, cosmonaut Alexey Leonov, one of the twenty Soviet Air Force pilots selected to join the original group of Soviet spacefarers, became the first man to walk in space. A trio of 16mm cameras—two in the airlock, one outside on a boom—recorded the remarkable first extravehicular activity (EVA). After training for eighteen months, the thirty-year-old Leonov found himself dangling outside his two-man Voskhod 2 spacecraft by a linguine-thin 17.6-foot umbilical cord about 310 miles above the Earth for twelve minutes and nine seconds of unmitigated misery. The strain of mankind's first EVA elevated his body temperature to heatstroke levels, while the vacuum bloated his space suit to Stay Puft Marshmallow Man dimensions.

At one point it became almost impossible for Leonov to move or even reenter the Voskhod 2's hatch. Leonov's labored (and recorded) respiration was so distressed it was later looped into the soundtrack of *2001: A Space Odyssey*, presumably when David Bowman forces his way into the airlock, dons a space helmet, and makes his way to the mainframe room to disconnect an anxious and beseeching HAL 9000. If things had gotten really rough, rumors persist that Leonov had a suicide pill tucked away in the event he was stranded outside the capsule for good.

When it came to outer space back then, the Americans were constantly pressing their noses against the window of greatness, copycatting every step the Russians took. A few weeks later, on June 3, astronaut Ed White became the first American to walk in space during his Gemini 4 mission. Starting over Hawaii and finishing over the Gulf of Mexico, White floated in an inky infinity for twenty-three minutes, nearly twice as long as Leonov. The jaunt released in him the kind of rapture that

Ed White, the first American to walk in space: "the saddest moment of my life." NASA

Owsley Stanley III was distilling into tablet form then. "I'm coming back in," he said of his walk, "and it's the saddest moment of my life."

During his high-flying, if heartbreaking, stroll, White propelled himself back and forth to the spacecraft three times using an oxygen-jet-powered "gun," more formally known as a Hand-Held Self-Maneuvering Unit. After he exhausted the gun's fuel in the first three minutes, White maneuvered by a combination of contorting his body for torque and yanking on the 23-foot tether line and the 5-foot umbilical line, both enclosed in gold tape that formed one cord. While White's walk was longer and

White floats out the open hatch. NASA

happier than Leonov's, it nevertheless mirrored the Russian's experience in one crucial way: If White had passed out from lack of oxygen, his co-pilot inside the spacecraft, James McDivitt, was under orders to sever his cord, because NASA, in the survival calculus of space exploration, thought trying to pull him back through the hatch would risk two lives instead of just one.

NASA's ambitions for the EVAs were greater than having its astronauts floating like balloons in the Macy's Thanksgiving Day parade. The agency needed to find a way for them to maneuver efficiently through the vacuum, zero-, and microgravity of space while also conducting physically difficult, mentally complicated tasks that involved, say, digging or hammering. Merely twisting a wrench with a bit too much elbow grease, for example, might result in a Newtonian horror show that could fling an astronaut to infinity and beyond.

The oxygen jet gun, not to mention panicked flailing, were neither precise nor powerful enough to be useful. NASA's solution to the

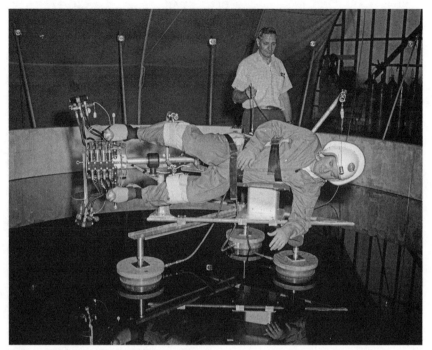

The jet shoes simulator. NASA

problem stretched back to the last days of World War II, when Charles Zimmerman developed his "Flying Shoes," which consisted of two small four-cylinder target drone engines powering upward-facing propellers. Not unlike a Segway today, a pilot would buckle into its metal bindings and steer in any direction by shifting his weight forward to go straight ahead, for example, or leaning to the left to go that direction, or bearing right to go that way.

The Flying Shoes morphed into a "Flying Platform" by Zimmerman and colleague Paul Hill. First conceived as a military tool, then repurposed conceptually as a lunar explorer, it was a circular podium powered by ducted fans positioned underneath its pilot. Early test runs achieved altitudes up to 12 feet and speeds up to 17 knots.

Neither the Flying Platform nor the Flying Shoes got off the ground, but Bird found inspiration in them nonetheless. His jet shoe concept included single jets placed on the soles of the astronaut's shoes.

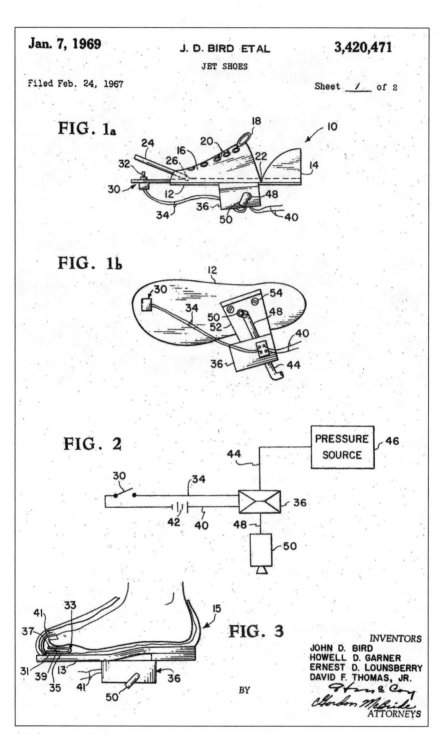

FIG. 1a

FIG. 1b

FIG. 2

PRESSURE SOURCE

FIG. 3

INVENTORS
JOHN D. BIRD
HOWELL D. GARNER
ERNEST D. LOUNSBERRY
DAVID F. THOMAS, JR.

BY

ATTORNEYS

Jet shoes patent. NASA

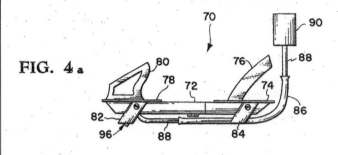

FIG. 4a

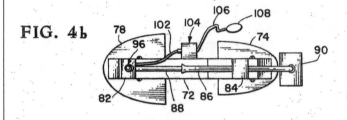

FIG. 4b

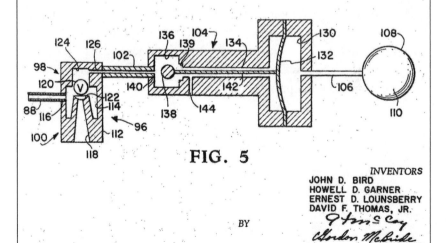

FIG. 5

INVENTORS
JOHN D. BIRD
HOWELL D. GARNER
ERNEST D. LOUNSBERRY
DAVID F. THOMAS, JR.

BY

ATTORNEYS

The astronaut would then activate the jets by stepping down on a switch with his big toe that would release a burst of pressurized gas. Then he would guide himself by using his feet and legs as a kind of bodily joystick for attitude and position. (Other potential starter/navigation included mouth-based controls, such as a bite bar, or a sip-switch.)

Early versions of the jet shoes were considered too heavy and clunky to be practical. Later, Bird constructed a version made of a pair of aluminum roller skates that could be readily attached to shoes or boots of any size by a spring system.

Bird's power shoes never got the traction he expected. On April 6, 1967, Deke Slayton, then director of Flight Crew Operations, asked for the jet shoes to be discontinued, reasoning that adding them to the already-burdensome space suit (including its life-support backpack, the Apollo suit weighed about 180 pounds) would create a safety risk.

It was an ending, but not the end. The idea of a portable, ergonomic propulsion system unit evolved into NASA's Manned Maneuvering Unit (MMU), with the much cooler nickname of "space jetpack." First tested in space in 1984 by astronaut Bruce McCandless, the space jetpack was tried out on nine sorties by a total of six astronauts.

Following the MMU came the Simplified Aid for EVA Rescue (SAFER) in 1994. A life jacket for spacewalks, SAFER was worn like a backpack and was powered by twenty-four small, fixed-position, nitrogen-jet thrusters that allowed an astronaut to maneuver around in space. Intended primarily for emergency rescues, SAFER didn't replace the stone-age tethers and safety grips still in use. Today, fifty-plus years after Ed White sailed a celestial sea, NASA is developing versions of the space jetpack that will aviate over asteroids and cruise over comets.

Iron Man would approve.

Chapter Six

Point and Shoot:
Ray Guns

FOR MANY PEOPLE, THE TRUE SOUND OF THE SPACE AGE IS NOT THE telegraph-like *beep-beep-beep* of Sputnik, the oceanic roar of a Saturn V rocket, or the even the scratchy sound of Neil Armstrong's eternal flame of a quotation.

It's the *pew-pew-pew* sound of ray guns.

Burrowed into that onomatopoetic noise is our passage to what the space program represented: a future far beyond the grubby limitations of diapered astronauts, fifteen-minute spaceflights, spaceships decorated in cramped, no-frills confines that shouted "early U-boat," muscles atrophying and bones weakening in weightlessness, and sweat and urine recycled for drinking water. It is difficult to get excited about a medium where even at a velocity of 10 percent the speed of light, it would take 42.4 years to reach the Proxima Centauri, the star closest to us. At the same speed, it would take 4,900 years to reach Kepler-186f, likely the most habitable of the exoplanets yet discovered.

What made those limitations bearable—and thus made continuing enthusiasm for the space program possible—was the wonder of things-that-could-be, like the ray gun. The space-age firearm could be carried like a Saturday Night Special, but its "bullets" of energy could take down a foe with the reliability of a Kalashnikov, and the fearsomeness of a flamethrower.

Ray guns let us feel like intergalactic cowboys on the high plains of Mars. They suggested not the Michelin Man bulk of space suits, or the drugged-sloth slow crawl of spaceships, but adventure as fast as light itself. Washington Irving, in his 1809 book *Knickerbocker's History of New York*, postulated an invasion by moon people, who he pictured as "riding on hippogriffs—defended with impenetrable armor—armed with concentrated sunbeams." But perhaps the true forefather of these atomizing artilleries was H. G. Wells and his "Heat-Ray" wielded by the conquering Martians in their tripod ships in *The War of the Worlds*:

> *Slowly a humped shape rose out of the pit, and the ghost of a beam of light seemed to flicker out from it.*
>
> *Forthwith flashes of actual flame, a bright glare leaping from one to another, sprang from the scattered group of men. It was as if some invisible jet impinged upon them and flashed into white flame. It was as if each man were suddenly and momentarily turned to fire.*

In Wells's version, the weapon was a kind of laser with a visible beam, while in Steven Spielberg's 2005 couture-friendly account, it only boiled people, saving their garments from heated dissolution.

Oddly enough, a subset of the ray gun, the disintegrator ray, appeared first in *Edison's Conquest of Mars*, by Garrett Putnam Serviss, an American astronomer, and popularizer of science. Serialized in early 1898 in the *New York Evening Journal* as "The Conquest of Mars," and later the same year, in the *Boston Post* under the full book title, it was a kind of Victorian fan-fiction crossover of Wells and the Wizard of Menlo Park, capitalizing on the widespread celebrity of both, one in fiction and one in fact. The book lays out a shocked-and-awed world following the aftermath of a major Martian attack. Savaged and stunned, the Earth decides to turn the cosmic tables and travel to the Red Planet to give its inhabitants a big, Gladiator-style thumbs-down.

> *Why should we wait? Why should we run the risk of having our cities destroyed and our lands desolated a second time? Let us go to Mars. We have the means. Let us beard the lion in his den. Let us ourselves turn conquerors and take possession of that detestable planet, and if necessary, destroy it in order to relieve the earth of this perpetual threat which now hangs over us like the sword of Damocles.*

Not unlike Jason and the Argonauts an epoch before, or *Armageddon* and *Independence Day* a century later, the story enlisted a team of super-science heroes—although some of these champions, including Lord Kelvin and Wilhelm Röntgen, were as real as Wells himself. President McKinley, Queen Victoria, Czar Nicholas, and the emperor of China also make cameo appearances. The book introduced a mélange of now-venerable tropes, including aliens building the pyramids, alien abductions, asteroid mining, space battles, functioning space suits, and, course, ray guns.

I had the good fortune to be present when this powerful engine of destruction was submitted to its first test. We had gone upon the roof of Mr. Edison's laboratory and the inventor held the little instrument, with its attached mirror, in his hand. We looked about for some object on which to try its powers. On a bare limb of a tree not far away, for it was late fall, sat a disconsolate crow.

"Good," said Mr. Edison, "that will do." He touched a button at the side of the instrument and a soft, whirring noise was heard. "Feathers," said Mr. Edison, "have a vibration period of three hundred and eighty-six million per second."

He adjusted the index as he spoke. Then, through a sighting tube, he aimed at the bird.

"Now watch," he said.

Another soft whirr in the instrument, a momentary flash of light close around it, and, behold, the crow had turned from black to white!

"Its feathers are gone," said the inventor; "they have been dissipated into their constituent atoms. Now, we will finish the crow."

Instantly there was another adjustment of the index, another outshooting of vibratory force, a rapid up and down motion of the index to include a certain range of vibrations, and the crow itself was gone—vanished in empty space! There was the bare twig on which a moment before it had stood. Behind, in the sky, was the white cloud against which its black form had been sharply outlined, but there was no more crow.

"That looks bad for the Martians, doesn't it?" said the Wizard.

While Serviss may not have posed a threat to knock Jules Verne out of the science-fiction hall of fame, he did have a gift for literary under-statement.

With Edison's aid, the armada of Earthmen washes the Martians off their planet home like a cheap watercolor. The book is now forgotten, but its legacy of the ray gun persists in phasers and blasters. Irving and Serviss had the correct concept if not the catchy nomenclature down. The first recorded instance of the term *ray gun* came courtesy of Victor Rousseau in *The Messiah of the Cylinder* in 1917. A few years later, Philip Francis Nowlan put a disintegrator into the hands of his character Buck Rogers, and E. E. "Doc" Smith hung ray guns on the proverbial wall in his Lensman novels.

The ray gun's cultural shelf life, however, might have been an abbre-viated one if not for a real-life version, or at least the believable report of it: By 1915 scientist Nikola Tesla, the onetime protégé of Thomas Edison, wrote in a letter to the editor of the *New York Times* that "It is perfectly practical to transmit electrical energy without wires and produce destruc-tive effects at a distance. I have already constructed a wireless transmitter which makes this possible . . . When unavoidable, the [transmitter] may be used to destroy property and life." A 1934 *New York Times* headline blared "Tesla, at 78, Bares New 'Death Beam,'" explaining that the lethal ray "will send concentrated beams of particles through the free air, of such tremendous energy that they will bring down a fleet of 10,000 enemy airplanes at a distance of 250 miles."

Tesla insisted war would then be made impossible. The beam—today we could likely call it a charged-particle-beam weapon—would offer every country that has one the opportunity to build an "invisible Chinese wall" around its borders, keeping if safe and secure.

He tried, and failed, to interest J. P. Morgan Jr. and Prime Minister Neville Chamberlain of Great Britain to finance his device. His letter to Morgan nudged him to pull out his wallet, since "[the] Russians are very anxious to render their borders safe against Japanese invasion and I have made them a proposal which is being seriously considered." Not surpris-ingly, Tesla's lemonade-stand sales tactic did not persuade Morgan to act.

By 1937, exasperated by his failed attempts to produce attention and bankrolling for his invention, Tesla wrote an intricate, technically dense paper, "New Art of Projecting Concentrated Non-Dispersive Energy through Natural Media." He sent the paper to a host of countries, including the United States, the Soviet Union, England, France, and the Kingdom of Yugoslavia.

The only nibbles came from Russia and Yugoslavia, both keenly aware of their vulnerability to the impending Nazi threat. Tesla presented his scheme in 1937 to the Amtorg Trading Corp. in New York, which handled the greater part of Soviet-American trade, and served as a cover for Soviet industrial espionage. (The company, founded in 1924, once offered a job to L. Ron Hubbard, the founder of Scientology.) Two years later, when one stage of the plan was supposedly tested to the USSR's satisfaction, Tesla received a check for $25,000.

The death beam begins to fade here. It crops up in the *Baltimore Sun* in July of 1940, as America's inescapable entry into the war drew near. With the shattering bombings of Guernica in 1937 and Rotterdam in 1940, the paper understandably highlighted the beam's theoretical ability to strike airline motors 250 miles away. The *Philadelphia Inquirer* takes note of it in October, 1940, but war—and Tesla's death in 1943, alone in his room at the Hotel New Yorker—dumped the inventor and his almost-clairvoyant concepts into obscurity for several years.

And, too, there was the continually disappointing reality when real-life energy weapons fell short of their promise. Even the invention of the world's first laser (its name stood for "Light Amplification by Stimulated Emission of Radiation), known as the "ruby laser," in 1960 was a letdown instead of a breakthrough when it came to quickly adapting it as a real-world ray gun. Even after years of development, several key issues had to be resolved: A weaponized laser beam would still require an enormous power pack to punch with the impact of a bullet; and then there were the laser's additional problems of recharging times, heat dissipation, and the troubling trend of its photons being deflected when they bump into water droplets in rain or fog.

China announced in 2018 that it had developed a 6.6-pound portable laser rifle. Designed by the Xian Institute of Optics and Precision Mechanics in Xian, Shaanxi, and made by ZKZM Laser, it can supposedly scar skin, set clothes on fire, and take down a drone from a maximum range of 2,625 feet. Its lithium-ion battery gives it enough energy to power one thousand two-second-long shots. Even if real, it would almost certainly violate the 1998 United Nations Protocol on Blinding Laser Weapons—especially given that it would be used to fire on domestic crowds instead of the Klingons.

Still the dream of space weapons lived on even though the reality of armed astronauts was more flintlock than phaser. The 1959 Camillus folding knife survival kit was carried aboard 1961's Mercury Freedom 7 as part of Alan Shepard's survival kit. Later Mercury astronauts were equipped with Model 17 "Astro" hand-wrought survival knives that sported a 5.5-inch blade tough enough to wedge open the capsule hatch should the occasion arise.

With the Gemini and Apollo missions intended to splash down near the equator, NASA commissioned a survival knife, the M-1, fit for jungle survival. Made by W. R. Case & Sons Cutlery Co., the M-1 came with a lightweight polypropylene handle, a 17-inch blade, and a ridge of saw teeth along the blade's back. The rest of the kit—like the one astronauts Jim McDivitt and Ed White carried on Gemini IV—might include a first-aid kit, survival radio, signaling mirror, shark repellent, seawater desalting tablets, sunscreen, soap, and sunglasses.

Not surprising for a country that once invented a flying tank, the Soviets had cooler weapons. For decades, cosmonauts on Soyuz carried a TP-82 gun, a specially modified weapon with three barrels for three types of ammunition—shotgun shells, rifle bullets, and flares—that were carried in a belt attached to the gun. The TP-82's folding stock doubled as a shovel and contained a swing-out machete.

The Russians came loaded for bear—literally. Since the Soviet ships landed on ground instead of water, the weapon would protect the cosmonauts in case their spacecraft wandered far from their intended touchdown points, and ended up in ursine-rich Siberia. (Russia has the largest black bear population in the world.) Besides the guns, the Soviet survival kit contained food rations, water bottles, fishhooks, warm clothing, and rope for making a shelter out of the Soyuz capsule's parachute.

Unlike the American astronauts, the cosmonauts actually ended up needing their survival kits. When Alexey Leonov and Pavel Belyayev returned with their Voskhod 2 spacecraft on March 18, 1965, from the first spacewalk, they ended up almost 1,250 miles from their planned landing site in the bear-filled Siberian forest. Surrounded by walls of snow more than 6 feet high, Leonov and Belyayev sat cold, hungry, marooned. At first helicopters could only manage to land about 3 miles away, in the meantime dropping warm clothes and supplies for the trapped two. In time a rescue team reached them, but even then, cosmonauts and rescuers had to ski their way out to the nearest landing zone.

Starting in the 1980s, the Soviets also brought lasers to what would have been a survival-knife fight. Before the Soviet Union fell, it developed handheld laser weapons. Instead of an OK-Corral-in-low-orbit gunfight, the weapon would likely have used pyrotechnic flashbulb ammunition to disable optical sensors on satellites or spacecraft. Some accounts assert that the heat of the energy beams were powerful enough to burn through a helmet visor, or blind someone from 65 feet away.

Even putting aside the disheartening lack of Klingons and Romulans to zap, an astronaut's workday consisted of sitting around in awkward positions, excreting in a kind of jerry-rigged diaper, and eating mush out of tubes. Going into space, it turned out, was like going into a nursing home. Their "weapons" were equally a disappointing weak sauce. But the space voyages of fiction offered a starry realm where those commonplace limitations were irrelevant, and where the spaceman was naught without his ray gun, just as the cowboy was nothing without his six-shooter. Robert Heinlein's novels—1949's *Red Planet*, 1953's *Starman Jones*, and 1958's *Have Space Suit, Will Travel*—plus the sci-fi shows of the era— *Commando Cody: Sky Marshal of the Universe*, *Rocky Jones, Space Ranger*, and *Tom Corbett, Space Cadet*—mixed gunplay with galactic adventure. In a case of cultural CPR, toy ray guns kept the fantasy of a cosmos spilling over with adventure alive. The Tom Corbett Space Cadet Atomic Pistol from 1952 (rebranded a year later as the Official Rex Mars Planet Patrol Atomic) was sold by Louis Marx & Co. It was as iconic as a Colt or Remington, coming in Pacific-Ocean blue, police-siren red, and absinthe green. Though the Strato Gun cap gun and the Pyrotomic Disintegrator Pistol were products of the 1950s, they retained the exaggerated curves, kitschy colors, and sleek forms of Art Deco, which made them the futuristic version of the elegant showpieces decorated with precious jewels that Tiffany & Co. developed for Smith & Wesson handguns and Winchester rifles in the nineteenth century. Even if you shot your eye out, that unforgettable ray gun sound you made through puckered lips made it all worthwhile.

CHAPTER SEVEN

Drinking It In:
Liquid Refreshment

NASA DIDN'T INVENT TANG, BUT IT SHOULD HAVE. (IT DIDN'T INVENT Velcro or Teflon, either, though the myths about NASA's role in those will probably endure until the heat death of the universe.) The orange additive that gives mere tap water high-falutin' airs is as linked to the space program as donuts are with Homer Simpson.

Tang came to the world's attention on February 20, 1962, at a blistering pace of 17,000 miles per hour, nearly ten times the average speed of a bullet, when astronaut John Glenn was in the middle of his five-hour excursion around the Earth in Friendship 7. Thanks to an ongoing chemical reaction of the Mercury program's onboard life-support system, the water acquired an off-putting, if harmless, metallic taste that rendered the life-sustaining liquid not particularly attractive. To alleviate the slightly repellent flavor of the offensive H_2O, Glenn used the orange-flavored sugar powder known as Tang. With that simple gesture, Americans now had yet another iconic food brand to stand alongside their Coca-Cola and Twinkies.

Tang's journey into space began half a decade before Glenn's record-setting flight. In 1957, a food chemist named William Mitchell at General Foods Corporation invented what he called "Tang Flavor Crystals." Mitchell was the Thomas Edison of food/not-food, with more than seventy patents to his name for various nourishments more or less fit for human consumption. He was the controlling mind (and discerning tongue) behind such mid-century sustenance as Pop Rocks, Cool Whip, powdered egg whites, and a quick-set version of Jell-O.

After General Foods hired him as a research chemist in 1941, Mitchell developed a tapioca substitute. The main source for tapioca, the cassava plant, was mostly grown in the Far East, which meant it was to be under the control of the Axis powers. American troops referred to the scientist's standby replacement as "Mitchell's Mud," which sounds more like a Gordon Ramsay insult than an appreciation of wartime ingenuity.

Some of his lesser-known patents include a powder that when water was added would simulate the fizzy pleasure of a carbonated soda. He also patented ideas for a process to make carbonated ice. (Americans loved their gas back then.) Another of his concoctions, a coffee substitute, was whipped up from dahlias, whose roots are brown carrot-like tubers. Mitchell began marketing the faux stimulant as Dacopa, a coffee alternative loaded with health benefits. While it still survives today, Dacopa had no hope of usurping joltin' joe from its long-established throne.

After a period of research and development that lasted two full years, in 1959 General Foods placed Tang on grocery shelves throughout the United States, Venezuela, and West Germany, marketing the sugar powder not on its taste but on its logistical convenience: It was a breakfast drink that "you don't squeeze, unfreeze, or refrigerate." Even in an era of long-shelf-life products like Cheez Whiz, Sweet'N Low, Swanson TV dinners, and recipes such as tuna and Jell-O pie, tomato-banana tarts, beer and kraut fudge cake, Tang was a taste-bud outlier. Its reception was tepid; its sales were flat.

Some of the world's most revered brands require a tipping point to achieve an enduring fame and everlasting fortune. The Aunt Jemima brand, for example, was an edible nonentity until its makers exhibited it at the 1893 World's Columbian Exposition in Chicago. There they peddled fifty thousand of the pancake mixes, and within ten years of the world's fair became one of the most trusted brands in the country.

Tang needed an event of that size and scope, one that would take it from sinking ship to sensation. Glenn was in no danger of going Donner party. (He was alone, for one thing. Two-man flights didn't begin until the Gemini program.) But his triple-orbit trip around the Earth allowed him time to experiment with eating various foods. More important than what he ate, though, was finding out the most efficient ways to package and deliver food, and ascertaining whether humans can chew, swallow, and digest comfortably in zero gravity.

Glenn snacked on applesauce and then pureed beef and vegetable by squeezing them from an aluminum tube, like toothpaste. Though he decided to skip the tube of spaghetti, Glenn's culinary experience proved conclusively that peristalsis—the wave-like muscle contractions that move food through our digestive and urinary tracts—worked just as well in the much-reduced gravity of space as it did on Earth. He also proved the particular food-delivery system (i.e., the tubes), to be a competent means of snacking in space. Had peristalsis not worked, there was no plan B.

Running a distant second to anything the Soviets did in space back then was almost a given for the Americans, and that includes eating. Glenn wasn't the first to nosh in outer space. That laurel belonged to Yuri Gagarin in his Vostok spacecraft in 1961. Like Glenn, the first human in space ate from tubes, too, except his were filled with pureed meat and chocolate sauce.

The honor of being the first man to upchuck his meal in space also goes to a Russian, Gherman Titov. After downing two meals in his 1961 flight aboard Vostok 2, Titov, the second human to orbit the Earth, vomited it all back up. At first the Russians worried he might be genuinely ill, but later it turned out he was suffering only from "space sickness."

Today space sickness is more formally called space adaptation syndrome (SAS). Similar to motion sickness, SAS's symptoms run the gamut, from reduced appetite, gastrointestinal distress, and vertigo to nausea and vomiting. It's caused by the body adapting to prolonged weightlessness, especially the vestibular system, the apparatus in the ear that helps to regulate motion, equilibrium, and spatial orientation. As a rule of thumb, three-quarters of space travelers are predisposed to space sickness (which does not bode well for the future of space tourism). Titov's nausea and "space sickness" were more than understandable At age twenty-five, the youngest person to fly in space, Titov racked up seventeen and a half orbits, a distance of nearly 430,000 miles in all—nearly equal to a round-trip from the Earth to the moon. His voyage stretched the human body's tolerance for weightlessness more than any other had at the time. Too late for Titov, but fortunately for everyone else, trans-dermal dimenhydrinate antinausea patches are now used to reduce the effects of SAS as the body adapts to its newfound buoyancy.

Eating was one thing; drinking was another matter entirely. At some point around 1960, NASA realized the Tang drink powder might provide vitamin-enhanced refreshment in space and provide a palpable workaround for the whiffy water. Buying a quantity in bulk from General Foods, NASA suggested that "Tang" never be used. They decided to eliminate it from any packaging or references, using only "orange drink" or "xylose sugar tablets" instead

In space, nothing is easy. Everything is complicated—and potentially deadly. Even Tang. On Earth, you would scoop up a tablespoon of the carroty dust out of its jar and dump it into an open glass of water. In space, the simple acts of scooping and dumping could disperse the crystallized powder into the cabin, flinging its grainy particles into the thousand gaps and spaces offered by the electronics.

To get the drink from the pouch to Glenn's mouth, NASA designed a delivery method Rube Goldberg would have envied. First the astronaut had to squirt a water-filled needle into a vacuum-sealed bag that held the powder. Then, once he shook the pouch to disperse the Tang effectively, he poked a straw inside and sipped the concoction. It was, in other words, our first juice box in space.

Even though NASA never officially labeled the product, on board Glenn's flight or any other, that was irrelevant. Word nonetheless spread around the world almost before Glenn landed. General Foods gushed proudly in print and TV ads that it was chosen by the Gemini astronauts because it was packed with vitamins, and tasted great.

There are no second acts in American life, F. Scott Fitzgerald famously declared in his notes for the unfinished novel, *The Last Tycoon*. But there was another early version of that same quote that appeared in his 1932 essay, "My Lost City": "I once thought that there were no second acts in American lives, but there was certainly to be a second act to New York's boom days."

Because of John Glenn and the skunky water, Tang was resurrected, and reborn in a second act that would endure for decades. Around 1965 footage started appearing in Tang commercials with Gemini astronauts drinking it on their missions, with similar imagery appearing in print ads. Tang sponsored ABC's coverage of Apollo 8, the first manned flight around the moon. A year later, Tang launched its "For Spacemen and Earth Families" ad campaign.

One TV spot dating from this time presents a "dinner of the future with food in plastic pouches," harking back to the scientifically diminished food of *The Jetsons*. Panning over the spread of plastic-shrouded food, the voice-over explains, "This is a typical meal served to astronauts aboard Apollo spaceflights: oatmeal, sausage, toast, applesauce, and, in a special zero-gravity pouch, Tang." Adam and Eve. Calvin and Hobbes. White Russians and Jeff Lebowski. Add to these eternal pairings NASA and Tang.

Attaching themselves to the space program with the tenacity of a sandbur, General Foods, and its ad agency, Young and Rubicam, traveled to Houston to review ad layouts with NASA, coaxing and wheedling them to say out loud that astronauts drank Tang, or at least release a picture of them doing so. NASA always resisted, but in the end it probably didn't matter. This association might be considered an example of the Mandela Effect, where we collectively misremember facts or events in the same way—e.g., that South African president Nelson Mandela died in prison during the 1980s, when he actually passed away in 2013, or that Darth Vader says, "Luke, I am your father," when in reality he tells young Skywalker, "No, I am your father."

What might have been only fifteen minutes of fame for another brand ended up being almost fifty years of renown for Tang. When Homer Simpson couldn't get Tang on the 1994 episode "Deep Space Homer," he phones the highest authority in the land to protest. "Hello, is this President Clinton? Good. I figured if anyone knows how to get Tang, it'd be you." When a longtime castaway in 2017's *Kong: Skull Island* learns from a newly arrived group that men have landed on the moon while he's been stranded, he asks the visitors, "What do they eat up there?" To which one replies, "Tang."

By 2011, Tang, then owned by Kraft Foods, had become the conglomerate's twelfth billion-dollar brand, with its global sales nearly doubling since 2006. Adapting to a world racing to undo climate change, just as it did to one racing to reach the moon, Kraft played up the idea that Tang, needing only powder and tap water, does far less damage to the environment than bottled sodas and sports drinks.

It's now owned by a Kraft Foods spin-off, Mondelēz International Inc. Depending on where you buy it in the world today, you can get Tang in flavors like cashew, guava, horchata, and Jamaica hibiscus. Tang is now a cultural fixture from Indonesia to Malaysia to the Middle East, with sales surging more than 20 percent in the UAE and Saudi Arabia during Ramadan. Consequently, Mondelēz is now marketing Tang to India's Muslims as the ideal beverage after the traditional daylong fasts in Ramadan.

Not to be outdone by a mere powder, in the 1980s, Coca-Cola wanted its titular product to be the first soft drink consumed in outer space. To accomplish that modest goal, the company had engineered a special can that maintained the cola's bubbly effervescence without shooting out of the can, as soda has a habit of doing in weightlessness. The $250,000, 12-ounce space can used a special nozzle and valve switch to deploy the soda. Coca-Cola also wanted to test if spaceflight might alter the taste perception of its product.

When PepsiCo Inc. heard about it, the company, as fierce in its rivalry with Coke as the Americans were with the Russians, insisted its sugar water be allowed on board as well. NASA granted its approval and the company hastily engineered its own canister, which resembled a can of shaving cream. It also used a dispensing valve, and cost $14 million to design and produce.

Together, the two soft-drink giants' cans were formally christened the Carbonated Beverage Dispenser Evaluation payload soda. In 1985 four cans of Coke and four of Pepsi rode aboard the space shuttle Challenger on its STS-51F mission. The astronauts aboard the Challenger tested the drinks with scientific rigor and objectivity, with the day shift drinking Coke and the night shift guzzling Pepsi. The hoped-for cola smack-down never occurred, however. The lack of proper refrigeration and the effects of microgravity flattened both sodas to the point that neither company could claim victory.

CHAPTER EIGHT

Eating Out: Food

THERE IS NO CHIPOTLE IN SPACE.

Let the horror of that sink in for a minute. Given that astronauts need about 3,200 calories a day, and the Centers for Disease Control and Prevention report that US adults get an estimated 11.3 percent of their calories from fast food, the lack of access to a Five Guys burger, a Dairy Queen Blizzard, or Chick-fil-A waffle fries begs the question of where astronauts will get enough food to function on. Space voyages could end up in danger, with missions to Jupiter resembling Jamestown's Starving Time.

Food in space poses several unique hurdles that earlier astronauts were able to avoid simply because of the brevity of their extraterrestrial excursions. Alan Shepard's fifteen-minute-long flight and John Glenn's four-hour-and-fifty-six-minute-long orbits weren't long enough for either of them to experience genuine food insecurity.

The problem, of course, is the quadruple threat of time, speed, space, and power. The fastest craft so far includes the 2006 New Horizons mission to Pluto, which reached a speedometer-busting 100,000 miles per hour. (The 2018 Parker Solar Probe should hit 430,000 mph, about 250 times faster than a speeding bullet.) This means interplanetary trips, even at these speeds or greater, would require weeks instead of days or hours. With propulsive systems advancing as slowly as sloths, that means NASA and the private space companies still need to find a way to accommodate astronauts' appetites.

The problem of food in space is, if not literally rocket science, close enough to it. The slower the ship, the longer the trip, which means, on a long voyage to Mars, say, the sheer bulk of stored food would result in tons of added weight. Meanwhile, the vast electrical suck of refrigeration systems to preserve perishables is a technical hurdle well beyond the capacities of Rayovac and Eveready to solve.

The importance of food in space travel is at least as old as Jules Verne. In his 1865 novel *From the Earth to the Moon* (*De la Terre à la Lune*), the granddad of science fiction imagined the journey to the Earth's sole natural satellite punctuated by repasts worthy of a Michelin star. Gas-powered cooking allowed the crew's Frenchman (who was, naturally, declared chief cook) to whip up soups, and beefsteaks "compressed by a hydraulic press, as tender and succulent as if brought straight from the kitchen of an English eating-house." Preserved vegetables "fresher than nature" followed, then tea made from leaves gifted to the voyagers by the emperor of Russia. Topping it off was a fine wine distilled on the slopes of Burgundy.

Such an off-world buffet might have been Verne's greatest leap of speculative imagination ever. Most visions of food and the future—and, by implication, space travel—seem to imagine it concentrated into a form whose relationship to the original fare is no less unpleasant than that of a shrunken head to its original format.

Food packed within the confines of a pill had long been a techno-gourmand's fantasy since at least 1893. Anticipating the wonders that would be on prominent display to the millions attending the World's Columbian Exposition in Chicago—aka, the World's Fair—the American Press Association asked Americans to offer their wishful thinking / informed prophecies of the future. Mary Elizabeth Lease, a suffragist feminist from Kansas, foresaw a day when a food pill would liberate women from the domestic drudgery of cooking. "A small phial of this life from the fertile bosom of mother earth," she wrote, "will furnish man with subsistence for days, and thus the problems of cooks and cooking will be solved."

She wasn't alone by any means. L. Frank Baum, the author of the Wizard of Oz series, also believed in the magic of shriveled sustenance. From *The Patchwork Girl of Oz*:

He took a bottle from his pocket and shook from it a tablet about the size of one of Ojo's finger-nails.

"That," announced the Shaggy Man, "is a square meal, in condensed form. Invention of the great Professor Wogglebug of the Royal College of Athletics. It contains soup, fish, roast meat, salad, apple-dumplings, ice cream and chocolate-drops, all boiled down to this small size, so it can be conveniently carried and swallowed when you are hungry and need a square meal."

"John Jones's Dollar," a short story by H. S. Keeler, published in the August 1915 issue of *The Black Cat*, offered up elements of synthetic food tablets. So did the 1930 science-fiction musical, *Just Imagine*, where a man wakes from a fifty-year-long coma to find himself in ultramodern 1980s New York. At a café, he chows down on a "meal of clam chowder, roast beef, beets, asparagus, pie and coffee"—all packed neatly into a pill.

Much of the enthusiasm for food in emaciated form, leached of all visual appeal and palate-pleasing traits, likely stemmed from the food chain of nineteenth- and early-twentieth-century America possessing more poison than a Borgia family dinner.

Harvey Wiley, who became the chief chemist of the US Department of Agriculture in 1883, formed "poison squads" who investigated the mass contamination of the country's foods in that gastronomic Dark Age before the Pure Food and Drug Act of 1906. Wiley and his teams found a vast number of adulterated foods: Red lead was use to dye cheese a healthy orange pallor. Pork and beans and milk contained formaldehyde. Butter often contained borax, commonly used to kill ants. Beer was spiked with strychnine, and bread laced with chalk.

Still, hygienic food pills never mustered an allure past that resembling the better Jelly Belly jelly beans. *The Jetsons'* Dial-a-Meal that issued a burnt-toast pill, and *Star Trek: The Original Series'* food synthesizer that cooked up Play-Doh-like cubes of mystery meat probably consumed any desire by the general public for food that resembled pebbles more than pasta, and gravel more than gravy.

If fiction inadvertently staggered the food-pill idea, then fact dealt it a head butt. An adult generally requires at least 2,000 calories per day, and even the most concentrated source can't deliver that amount.

The food pill may have gone the way of other outmoded recurring themes in science fiction, like super-smart computers that detonate when the hero confuses them, or the myth that humans only use 10 percent of their brain. Science fiction didn't send the idea back to the kitchen, exactly. *Blade Runner* kicked off a trope of eating noodles in a ramen-dominated future that has since infected *Cowboy Bebop*, *The Fifth Element*, *Battlestar Galactica*, and *Almost Human*. A bland mush was the daily special in *Brazil*, and "people power" took on a new meaning in *Soylent Green*. These tropes' power endured because they married the idea that nature's food can be replaced with an improved version that is leaps and bounds more nourishing and convenient.

Like the space suits whose origins lay in military flight gear, astronaut food in the Project Mercury and Gemini program flights resembled armed forces' survival rations. John Glenn, flying the Friendship 7 in 1962, was the first American astronaut to nosh in space. He slurped pureed beef and vegetables and applesauce out of a tube and drank water with xylose sugar tablets (i.e., Tang). Glenn showed that people could chew, swallow, and digest food in a zero-gravity environment, something that was a matter of debate at the time.

For the longer-duration Gemini flights, NASA reasoned that in the weightlessness of space, astronauts would be less likely to expend the same amount of energy as they would doing similar work on Earth. So instead of packing the meals with 3,000 or so calories' worth of body fuel, NASA Gemini astronauts slimmed them down to 2,500 calories' worth instead.

The menu for Gemini was prepared and packaged by NASA, the Whirlpool Corp., and the US Army Laboratory in Natick, Massachusetts. It included dehydrated, freeze-dried, and bite-size foods. Since the fuel cells on board the Gemini craft manufactured water as a by-product, NASA particularly liked the idea of dehydrated foods. Such foods could be rehydrated later, in flight, thus saving critical weight on launch, when even a few ounces could make all the difference between a liftoff and a crash landing.

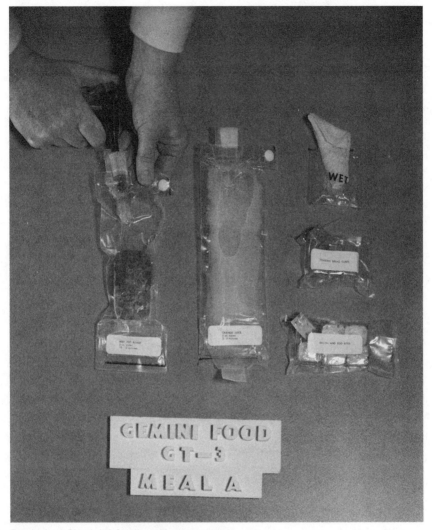

Prepacked Gemini 4 foods ready to go. NASA

For the first long-term flight—the fourteen-day Gemini 7 mission, with crew—the nutritional criteria was balanced precariously because of the capsule's weight limitations. The astronauts were each allotted about 1.7 pounds of food daily, stuffed into 110 cubic inches of packaging. Each Gemini meal came with four to seven servings of food in bite-size form (as compressed cubes, or in freeze-dried rectangular shapes) that could be munched on directly. The repasts included a very basic food-pyramid-approved selection of meats, bread, and dessert. The rehydratable foods included dry mixes and freeze-dried products, such as puddings, soups, entrees, fruits, and vegetables. The three-meal-per-day diet was designed to provide 16 to 17 percent of total calories from protein, 30 to 32 percent from fat, and 50 to 54 percent from carbohydrates.

Entertainment set in outer space commonly features deadly threats of asteroids, comets, and radiation, to name a few. Few find life-or-death drama in one of the most feared dangers of all: crumbs.

If NASA had a big red panic button in Mission Control, its officials would have slapped it during the 1965 Gemini 3 flight. While John Young and crewmate Gus Grissom were circling the planet, Young casually pulled out a corned-beef sandwich on rye and handed it to his crewmate.

The offending snack came from a Cocoa Beach, Florida, deli called Wolfie's Restaurant and Sandwich Shop. Wally Schirra had bought the sandwich and handed it off to Young, who snuck the contraband on board, concealed in one of his space suit's pockets. The mission transcripts reveal that a bemused Grissom took a few bites but didn't finish the surprise entree, because each piece he bit off started producing crumbs.

Grissom: What is it?

Young: Corn[ed]-beef sandwich.

Grissom: Where did that come from?

Young: I brought it with me. Let's see how it tastes. Smells, doesn't it?

Grissom: Yes, it's breaking up. I'm going to stick it in my pocket.

Young: Is it?

Young: It was a thought, anyways.

Grissom: Yep.

Young: Not a very good one.

Food flight: beef roast, bread cubes, and bacon. NASA

Grissom: Pretty good, though, if it would just hold together.

Young: Want some chicken leg?

Upset that Grissom and Young had supposedly sabotaged the part of the mission where they were supposed to evaluate the approved space foods, including the new dehydrated packets, the House Appropriations Committee met with NASA officials to discuss the ramification of the corned-beef affair. "My thought is that . . . to have one of the astronauts slip a sandwich aboard the vehicle, frankly, is just a little bit disgusting," said Congressman George Shipley of Illinois.

Whatever the offenses of the outlaw sandwich, crumbs have been the bane of NASA's food preparation for decades. The dangers they posed were many. The lack of gravity meant they could float into any crevice anywhere, disrupting equipment, and clogging ventilation systems. To prevent their crumbling, the foods were often coated with gelatin or oil.

Moreover, NASA doused powders like salt in a watery format, and pepper in an oily one to eliminate the chance of their particles roaming free. This is why crackers, cookies, and bread are generally verboten in space. Non-crumbling tortillas are often a staple of choice.

The list of awkward foods in space is a long one. Carbonated beverages are frowned upon. In low gravity, the bubbles inside soda arbitrarily spread throughout the fluid, even after swallowing. That results in belches that are more of a liquid spew whose contents resemble acid reflux. Even toothpaste is problematic, since spitting while brushing might result in nomads of dribble eternally traveling the spaceship. The solution was a foamless "ingestible toothpaste" that NASA later put on the market in the early 1980s.

Young's prank led to a NASA inquiry of just what other items the astronauts were sneaking abroad the spacecraft. The sandwich, though, lives on in the Grissom Memorial Museum in Mitchell, Indiana, where a corned-beef sandwich, preserved in acrylic, honors the illicit repast.

When it came to Apollo missions, it was like upgrading from Arby's to Alinea. A mutual effort among the US Air Force Manned Orbiting Laboratory Program, the US Army Natick Laboratories, industry, and various universities, the ever-evolving menu had to address several recurring problems. Astronauts complained that meal preparation and consumption took too much time, and that the water for reconstituting the dehydrated foods was sour-tasting. Even in-flight nausea was attributed to the food. Despite NASA's efforts to expand the selections and improve their taste, a kind of temporary "anorexia" that lasted the duration of the flight was seen as another repeated difficulty.

The expanded food system for the first manned Apollo mission, Apollo 7, of ninety-six food items helped increase the amount the crews would eat, which resolved at least the low-key hunger strike some went on. The astronauts on the flight, Wally Schirra, Walter Cunningham, and Donn Eisele, supped on corn flakes, peaches, pot roast, spaghetti, and shrimp cocktail, to name a few items. (Ice cream was scheduled to be part of the fare, but it mysteriously went AWOL.) Of all the foods, it turned out bacon was the most appreciated. "Happiness," commented Cunningham, "is bacon squares for breakfast."

The next flight, Apollo 8, saw NASA begin to abandon its dependence on rehydratable foods. During the first lunar orbital mission, on Christmas Day, 1968, Frank Borman, Jim Lovell, and William Anders opened packages of thermo-stabilized turkey and gravy. The traditional holiday repast had been stuffed in what was known as a "wetpack," made from a laminate of polyester, aluminum foil, and polyolefin. Since these ready-to-eat meals retained much of their water content, the astronauts didn't have to pump in the repellent water to prepare them. Cleverly, the dampness in the foods made them stick to the spoons they were eaten with, a certifiable plus in an environment with little gravity where everything not nailed down can become a roving menace.

The menus for Apollos 7 through 10 mostly consisted of three categories that reeked more of food science than food: rehydratable, dry bite, and intermediate moisture bite. Fruit cocktail, for example, came under the rehydratable grouping, cinnamon bread cubes qualified as a dry bite, while gingerbread fell into the intermediate moisture bite class.

By the time Apollo 11 headed to the moon, the tubes of squishy meat and applesauce had been long scrapped. Neil Armstrong, Buzz Aldrin, and Michael Collins noshed on food stuffed in the spoon-bowl packets that opened via a plastic zipper. The trio had their pick of sausage patties, pork with scalloped potatoes, and chicken. The meals were rehydrated and warmed in their pouch (not unlike microwavable foods).

The first meal on the moon for the first men on it was bacon squares—along with peaches, sugar cookie cubes, and pineapple grapefruit drink. Aldrin's favorite meal on the mission was shrimp cocktail, while Armstrong was partial to spaghetti with meat sauce.

In the world where the most dangerous drinking game "is seeing how long I can go without coffee," there's no reason the Apollo 11 team should have been bereft of their daily required java. For the first time, coffee went into outer space, making the quarter-million-mile journey to the moon. A total of fifteen cups were brought on board in the form of freeze-dried packets filled with powdered coffee. The pick-me-up was reconstituted with the help of a pistol-like water probe that dispensed one half-ounce of hot or cold water each time the trigger button was squeezed. (Tang, so closely entwined with the space program from its early days, was noticeably absent from Apollo 11.) Aldrin liked his joe black, Collins added sugar, and Armstrong preferred it light and sweet. More recently, in 2015, the Italian firms Lavazza and Argotec developed an espresso machine, called ISSpresso, for the International Space Station (ISS).

The food aboard Apollo 11 represented the pinnacle of late-1960s technology, every bit as much as the lunar landing module or the space suits. But the technology and ingenuity didn't stop with the moon landing. As NASA moved on into the era of the space shuttle and the International Space Station, it pondered how best to feed crews that would no doubt become larger on missions that would certainly become longer.

These days, the ISS pantry stocks about two hundred different foods and drinks, including scrambled eggs, chicken teriyaki, and oatmeal with raisins. (It takes NASA about two years of researching and developing a freeze-dried or thermo-stabilized version of a given food before it can be shipped into space.) The Russians reportedly feast on jellied pike perch and goulash with buckwheat (a diet which purportedly wreaked havoc with the ISS toilets).

One of the unforeseen consequences of living in space is that swelling in astronauts' nasal passages blocks the passage of a food's odor molecules to their nasal receptors. This results in a reduction in their ability to taste what they're eating by as much as 70 percent. As a result Tabasco, wasabi, and Louisiana Hot Sauce have become prized commodities.

Decades ago in the 1980s, NASA and Martin Marietta Corp. teamed up to see if various strains of microalgae could serve as a nutritional Swiss army knife, offering food supply, oxygen source, and catalyst for waste disposal on extended space voyages. The plant-based food would become an essential life-support system on long-duration flights, requiring minimal space and energy.

During this research scientists discovered that microalgae offered value not just to people hungry on spaceships, but to humans here on Earth, in the form of a nutritional supplement. Driven by these possibilities, the researchers spun off from Martin Marietta, and in 1985 formed Martek Biosciences Corp., in Columbia, Maryland. The principal discovery Martek had made about microalgae was identifying a strain that produces Docosahexaenoic acid (DHA), one of the most vital omega-3 fatty acids, and a primary structural component of the human brain, cerebral cortex, skin, and retina. But as crucial as it is to adult health, it's even more critical during the first year of life, when the human brain nearly triples in size. Without enough DHA, learning ability, mental development, and visual acuity are derailed.

Martek engineered a nutritional supplement based on DHA, as well as patented a method for developing arachidonic acid, another fatty acid important to infant health. The supplements are now added into eggs and bread, milk, yogurt, orange juice, and cooking oil. More importantly, they appear in 99 percent of the infant formulas sold in the United States, as well as in the infant formulas in more than seventy-five additional countries.

CHAPTER NINE

Consuming Passion:
Ice Cream

OUTER SPACE HOLDS MYSTERIES OF SUCH SCALE AND COMPLEXITY that it would take a cosmic Sherlock Holmes to solve them. Universe-spanning puzzles such as: What does dark matter consist of? Where does dark energy come from? What's behind the vast radio waves that create what's known as the "space roar"? And last but not least, did astronauts ever actually get to eat ice cream?

For the Apollo 7 mission in 1968, NASA issued a PR release that included the crew's bill of fare for its ten-day flight. Scheduled to orbit the Earth 163 times—the crew would spend more time in space than all the Soviet flights combined up to that point—meant the Apollo 7 menu had to be varied as much as it had to be nutritious. In fact, it could have easily been mistaken for the early-bird special at a modern Applebee's, encompassing fruit cocktail, cinnamon toast, chicken salad, pot roast, and butterscotch pudding. Gingerbread, date fruitcake, tuna salad and apple-sauce made an appearance too. And right there, in black and white, listed for meal B (i.e., lunch) on days "2, 6, and 10" was "vanilla ice cream."

But why is there no evidence that ice cream ever made the long voyage to space? Why does no one remember it? When asked about the ice cream years afterward, Walt Cunningham, then eighty-two, and the only surviving member of the Apollo 7 mission at that point, said he'd never eaten it. Cunningham did remember the freeze-dried chocolate pudding, though. He liked it so much he would horse-trade his other rations with fellow astronauts Walter Schirra and Donn Eisele for their chocolate puddings. It must have been good, because it wasn't easy to eat. Like any food on board the Apollo missions, ingesting it was a cumbersome process. To reconstitute the freeze-dried pudding, Cunningham first had to add water to a plastic bag with a water probe that meted out one half-ounce of hot or cold water. Then he had to shake it up, and after the rehydrated pudding finally formed into a slurry, he would eat the brown slush right out of the bag.

But ice cream? No, he was pretty sure there was never any ice cream.

Was space ice cream another great NASA hoax? Like the time in 1980 when Charles Johnson, president of the International Flat Earth Research Society, revealed how Arthur C. Clarke had written the script for a staged moon landing in order to bankrupt a Soviet Union desperately trying to keep up? (Johnson also explained that the stars are about as far above us as San Francisco is from Boston.)

Or what about that time in 2000, when Pierre Kohler, a French scientific writer, leaked a "secret" NASA report, "Document 12-571-3570," in his book *The Final Mission*? The covert document details a program for space shuttle mission STS-75, designed to test the most effective zero-gravity sex reproduction methods. A computer simulation pared twenty positions down to what it decided were an optimum ten, which were then exhaustively tested via human guinea pigs on STS-75.

Among the erogenous experiments' research findings were:

An inflatable tunnel enclosing and pressing the partners together. The partners faced each other in the standard missionary posture. The tunnel enclosed the partners roughly from the knees to waist and pressed them together with an air pressure of approximately 0.01 standard atmospheres.

Once properly aroused, the uniform pressure obtained from the tunnel was sufficient to allow fairly normal marital relations, but getting aroused while in the tunnel was difficult, and once aroused outside the tunnel, getting in was difficult.

This problem made the entire approach largely unusable.

These were obvious fakes and frauds, even, regrettably, the one about the NASA Sutra. (Though Arthur C. Clarke did say, "Weightlessness will bring new forms of erotica. About time, too." His track record for predictions was always pretty good.)

But what about ice cream? Did astronauts eat it and just forget? Were they covering something up? The truth is out there.

The lore of World War II includes tales of derring-do from combat to cooking, including how American heavy-bomber crews attached pairs of 5-gallon cans containing milk and cream to their planes. Tailored with a small propeller that whirled the mixture while the planes flew, the cans and their contents froze into edible ice cream as the aircraft soared to chilly high altitudes. Other versions say similar cans with a wind-driven spinner connected to a mixing rotor were hooked up under each wing of a Vought F4U Corsair fighter aircraft, with a similar appetizing outcome.

That was probably as close as ice cream ever actually came to achieving orbit. Despite the mass popularity of the confection known as "space ice cream" or "astronaut ice cream," it was never really part of the official astronaut diet.

When NASA began looking for ways to feed its crews on upcoming flights that would be considerably longer than Alan Shepard's toe-in-the-water fifteen-minute fight in 1961, it knew it needed to devise ways to package traditional meals in nontraditional ways. Like any house or computer hard drive, the spacecraft were tremendously limited in their storage space on the earliest missions. Refrigerators were too bulky, too heavy, and too power-intensive to realistically use on board, so whatever food NASA chose, it had to be storable at room temperature. That meant dehydrating, freeze-drying, or heat-treating nourishment for long-term—or, at least, longer-term—storage. One of those foods would be ice cream, which by then was as red, white, and blue as the hot dog, peanut butter, or Spam.

Even as far back as 1843, when New Yorker Nancy Johnson applied for a patent for her hand-cranked ice-cream freezer, ice cream was already an established landmark on the country's dietary landscape. The country's first published ice-cream recipe appeared in *The New Art of Cookery*, in

1792. George Washington spent $200 on the chilled treat in 1790, at one point purchasing an ice-cream maker comprising two pewter bowls. Thomas Jefferson served ice cream at the White House, the first president to do so. (His recipe, one of only ten surviving in his handwriting, most likely dates to his time in France. It required eighteen steps, and is said to resemble in its completed form a baked Alaska.) Dolly Madison, the wife of President James Madison, supplied strawberry ice cream at her husband's second inaugural ball.

Jump ahead to the Prohibition years, when the Eighteenth Amendment gave a booster shot to more than just organized crime. The Anheuser-Busch, Stroh's, and Yuengling brands survived the thirteen-year drought in part by making ice cream, and insinuating it even more into the American dietary tract. By 1958, the year NASA was officially formed, Haagen-Dazs announced its "gourmet" ice cream, bragging that it had 12 percent butterfat instead of the standard 10. By 1960, Good Humor offered 85 different flavors or combinations of ice cream. Howard Johnson's, of the ubiquitous "28 flavors" of ice cream (buttercrunch was inarguably the best), even had a product placement in *2001: A Space Odyssey*: After the flight to the space station in a Pan Am ship, Dr. Heywood Floyd walks through a Hilton hotel lobby, calls his family on an AT&T videophone, and then strolls by a "Howard Johnson's Earthlight Room."

So by the time the space race was running at full speed, few would have disputed that ice cream belonged on a confectionary Mount Rushmore. Adding it to the astronaut menu was a no-brainer.

Given the confines of spacecraft and the limitations of the available technology, NASA researched various ways food might be better stored and served. Early flights, like John Glenn's historical orbits in Friendship 7, had stocked food stuffed into tubes, like a Crest version of an hors d'oeuvre. One early and promising avenue was the freeze-drying process, in which water is extracted from fresh foods by a superfast dehydration process.

NASA contracted the Whirlpool Corp. to figure out a way to store ice cream at room temperature. The company's engineers invented a way to remove the water from ice cream without melting it. First, after cooling the ice cream to 5 degrees Fahrenheit, they used a vacuum pump that evapo-

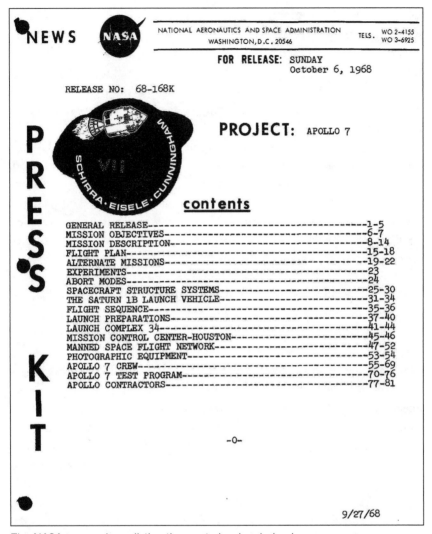

The NASA press release listing the mysteriously missing ice cream. NASA

rated the ice while still keeping the ice cream cold. They applied heat that turned its ice crystals directly from a solid state into a vapor, then captured and eradicated the vaporized water. The process repeats for hours, until the ice cream turns into a hard, easily crumbled slab. (Freeze-dried items also have to have enough water removed so that no dangerous bacteria can grow in them while they sit on the shelf.) Especially convenient for weight-sensitive spacecraft, the end product weighs only 20 percent what it did to start, but preserves 98 percent of its nutritional value. Then the ice cream was sealed in a pouch, where it can be kept at room temperature without melting. At that point, when you eat the converted product, the ice cream's brittle texture reverts slightly to its old consistency, which gives it an acceptably buttery feel.

Then NASA issued the fatal press release for Apollo 7. The 103-page document was a Wikipedia-long list of everything about the mission, from the experiments, such as shooting Earth photos with a 70 mm Hasselblad, to the abort modes, in case everything went to hell in space. Ice cream is indeed listed there in black and white, but no one recalled eating it. That matches up with the complete absence of any mention of ice cream from the mission's transcripts.

Most likely NASA grounded the ice cream late in the game, because it worried—rightly—that the food's brittle texture might produce an uncontrollable flurry of crumbs. Crumbs are a recurring concern any time food in space is discussed, a dread that shadows every nibble of every solid food. Dropped food wouldn't just descend to the floor where a friendly debate over the five-second rule might ensue. The weightlessness of space meant crumbs might float into the spacecraft's electronics and, like a malevolent army of scraps, wreak havoc.

Those with a sweet tooth and a jones for anything interplanetary can rejoice: Happily, space ice cream is sold at the US Space & Rocket Center in Huntsville, Alabama, and the National Air and Space Museum in Washington, DC. The flavors include Neapolitan and strawberry, and it even comes in an ice-cream sandwich version now, too.

Ice cream finally got off the ground and made it to space in 2006. Hauling a freezer to the International Space Station (ISS) for storing research samples, the space shuttle Discovery included several ice-cream cups from Blue Bell, a dairy in Texas near the Johnson Space Center. The ice cream was vanilla-flavored, with swirled-chocolate sauce.

Year after, SpaceX's CRS–12 2017 resupply mission to the ISS brought along thirty individual cups, again from Blue Bell. This time the haul consisted of tiny cups of vanilla, chocolate, and birthday-cake-flavored ice cream. Some Snickers ice-cream bars were included in the mix, too. (Brand logos aren't allowed on products going to the space station, so most are repackaged before launch to look as plain and generic as possible.)

No one should mourn the fact that space ice cream turned out to be two scoops of nothing. The undisputed king of sweets—chocolate—has been in space almost as often as Captain Kirk. The most enduring source of human confections, the cacao plant was cultivated thirty centuries ago in Mesoamerica. There, the cacao beans were so valuable a commodity, they served as ready currency. Chocolate has reigned as the king of sweets ever since, with no usurper in sight. Legend says Napoleon carried chocolate with him into battle. Marie Antoinette loved sipping hot chocolate. Spanish royalty included cakes of cacao in their dowries. Voltaire thought Columbus bringing back syphilis from the New World was more than offset by his also bringing back chocolate.

Astronauts treat chocolate with no less reverence than Napoleon or Voltaire. The first meal in space, Yuri Gagarin's, included chocolate sauce. They have eaten Dove Bars, Kit Kats, Tobler bars, and, of course, Milky Way bars while in orbit. In 1978, the Soviet cosmonauts on board Soyuz 29 carried chocolates with them into space, which got loose somehow and took them two hours to collect.

Even better, M&Ms turned out to be the perfect snack for space, precisely because their hard shell negates the chance of crumbs. Plus, their size makes them easy to pack, and simple to handle. Although M&Ms have been a part of every space shuttle mission since Columbia first launched in 1981, NASA, ever conscious of avoiding the slightest hint of endorsing any particular brand, refers to them as "candy-coated chocolates." That doesn't change the fact that whether on Earth or in space, as their slogan promises, they melt in your mouth, not in your hand.

CHAPTER TEN

Rage Against the Latrine: Diapers

DISPOSABLE ABSORBENT PADS FOR INFANTS EVOLVED AS QUICKLY AS America's postwar population baby-boomed. In 1946 Marion Donovan developed the "Boater," a disposable diaper that consisted of a conventional cloth diaper placed in a piece of shower-curtain plastic. Two years later, Johnson & Johnson introduced a mass-marketed disposable diaper—a pinless, snap-on version; and the Safe-T Di-Dee diaper appeared in 1950. Procter & Gamble Co. brought Pampers to the public in 1961.

Today, according to the Real Diaper Association, roughly 27 billion disposable diapers are used every year. "What happens in Huggies stays in Huggies," goes an ad slogan, suggesting parents could avoid messes that were once sopped up with milkweed leaf wraps, strips of linen, wool animal skins, or other natural fabrics.

Absolutely none of that helped Alan Shepard's weak bladder.

"Man, I gotta pee." As historic, *Bartlett's-Familiar-Quotations*-worthy, space-related quotes go, it wasn't quite on the same gilded pedestal as "That's one small step for man . . ." or "We came in peace for all mankind." Or even "Houston, we have a problem." But the words nonetheless captured with snapshot clarity the moment when biological urges as old as animal life itself almost derailed the 34-ton Mercury-Redstone rocket made of a special alloy from titanium and beryllium, known as René 41.

On May 5, 1961, Alan Shepard fidgeted while anxiously waiting inside his Freedom 7 capsule to become the first American in space. He was the aeronautical avatar for a United States desperately hurrying to catch up to the Soviets, who just three weeks before had sent cosmonaut

Prelaunch activity for Freedom 7. NASA

Yuri Gagarin aloft, helping him achieve the honor of being the first man in space in his Vostok spacecraft.

Early in the morning, Shepard began the laborious process of dressing for work, a process that in its length and complexity would rival a king dressing for his coronation. Four electrocardiograph pads were stuck to his chest, a respirometer tied to his neck, and because nothing is too undignified for science, a rectal thermometer slid into him to measure his body temperature. The first layer finished, he pulled on a set of long underwear and climbed into his shiny space suit, locking a multitude of zippers and connectors. By the time he was finished, Shepard was perspiring liberally and breathing laboriously. Shortly before the clock struck four in the morning, Shepard, accompanied by fellow astronaut Gus Grissom, rode in the transport van to the launchpad. There at 5:15 a.m., Shepard rose on an elevator to reach the 65-foot level (nicknamed "The Greenhouse"), which surrounded the capsule's hatch. Inside the capsule

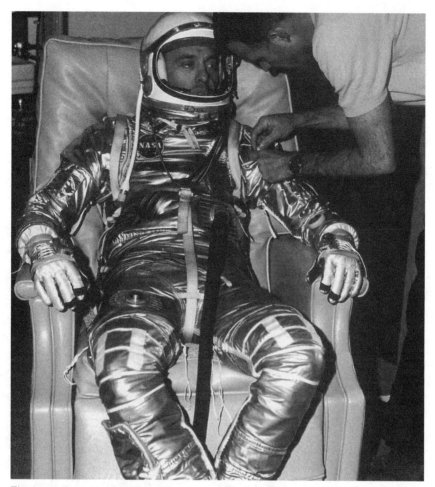

The long, slow prep for takeoff. NASA

he was helped into his specially contoured couch by Joseph Schmitt, the suit technician. Known better as the "valet of the space race," Schmitt secured Shepard, affixing straps across his chest, lap, knees, toes, and shoulders. At 6:10 a.m. the capsule's hatch closed with a deep thud.

Shepard, a space-age turtle on its back, looked around and laughed at what he saw.

John Glenn had put a girlie pinup along with a sign that read NO HANDBALL PLAYING IN THIS AREA. Shepard was good to go.

Then . . . nothing.

Space, like combat, is long stretches of boredom punctuated by short bursts of brain-freezing, gut-churning excitement. A stream of technical glitches turned into a flood: banks of clouds rolling in; a power invertor for the booster showing signs of trouble; a problem cropping up with one of the computers at the Goddard Space Flight Center. With all those delays, the onetime naval aviator Shepard stewed in the claustrophobic 7-foot-long, 6-foot-diameter capsule for so long, nature didn't so much call as screech into a megaphone.

Shepard pleaded his case to Gordon Cooper, another member of the original Mercury Seven astronauts, who was serving as the principal prelaunch communicator at Mission Control:

Shepard: Gordo!

Gordo: Go, Alan.

Shepard: Man, I got to pee.

Gordo: You what?

Shepard: You heard me. I've got to pee. I've been up here forever.

Forever in this instance meant nearly four hours. The idea of Shepard slipping out of his Mercury space suit—a custom-fitted, adapted version of the Goodrich US Navy Mark IV high-altitude jet-aircraft pressure suit—after all the time and trouble getting into it, was impractical. After much debate, the issue was kicked upstairs all the way to Wernher von Braun, the head of NASA's Marshall Space Flight Center in Huntsville, Alabama. Von Braun declared in his imperial way that "Ze astronaut shall stay in ze nosecone."

After all his hours of waiting patiently, all Shepard was asking for was a break that would require less than thirty seconds. We now know this because of the "law of urination." And we know the law of urination because, in 2013, intrepid researchers at the Georgia Institute of Technology proved, after filming a Noah's Ark's worth of elephants, goats, dogs, and rats in the Atlanta Zoo and the wild, that the time a mammal takes to empty a full bladder is proportional to the animal's mass raised to the power of a sixth. This means that mammals above a particular size, and more than 2.2 pounds, take about twenty-one seconds to empty their bladder. It's a medical fact.

Shepard threatened that he would urinate in his suit if he wasn't allowed to relieve himself the more-adult way, even though flight officials fretted that his letting go might short-circuit the medical wiring and electrical thermometers. Not to mention offending the estimated 45 million Americans watching or listening.

In the end, NASA decided to take the path of least resistance: They shut down the monitoring system and Shepard was given the green light to pee, saying "Do it in the suit!"

Shepard relieved himself with a denouement flourish of "Ahhhhhh," then declared, in the humor of the time, "I'm a wetback now," as the warm fluid pooled in the small of his back. His long cotton underwear absorbed the urine, which rapidly evaporated due to the atmosphere of 100 percent pure oxygen.

Because Shepard's flight was just fifteen minutes, in which he reached an altitude of over 116 miles and a maximum speed of 5,134 miles per hour, no one had planned to equip him with space diapers to accommodate any distracting physiological needs. But the need for a high-altitude bathroom break loomed large. John Glenn's bladder was proof of that.

When Glenn flew aboard Friendship 7, he excreted a firehose of 27 ounces of fluid in single space pee, 7 ounces more than the capacity of the average human bladder holds. Why so much?

In regular gravity the urge to urinate starts when the human bladder is only one-third full, because nerves covering the bladder sense it at that level. In zero or low gravity, the red alert to urinate doesn't begin until the bladder is almost completely full. By then, when you go, you go with a tank ready to slosh over.

It would only go downhill from there. NASA recognized that the very nature of spaceflight—even before astronauts were in the air—increased the need to urinate. Before and during takeoff, astronauts were strapped into chairs with their knees and legs raised above their heads, a position that naturally intensifies the need to urinate—so much so that

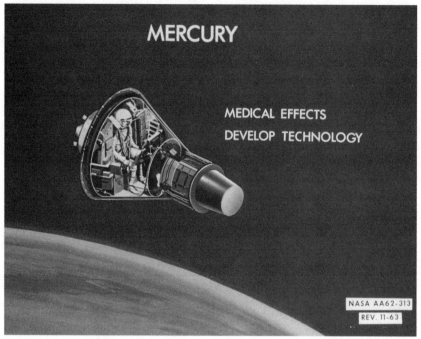

MERCURY

MEDICAL EFFECTS
DEVELOP TECHNOLOGY

NASA AA62-313
REV. 11-63

An artist's concept of the Mercury program study of medical effects and technology.
NASA

NASA estimated astronauts' kidneys would eject about a liter of urine while trapped in the launch position.

Worse for NASA, and much worse for the astronauts, upcoming missions would extend several hours, and, much later, trips to the moon would eventually require days. Even spacewalks, which lasted ten and twenty-three minutes the first times when Alexey Leonov and Ed White took them, respectively, now last between five and eight hours outside the International Space Station.

Long before Shepard's and Glenn's historic micturitions, military efforts were already well under way to concoct a urinary-collection device for flights of extreme duration, like those in the U-2 program, which took place at 70,000 feet, often for nine hours at a time. Pilots had to guide the spy plane, snap pictures of enemy territory, and sometimes evade those enemies' attempts to bring them down with a scrum of surface-to-air missiles. Not the kind of job where you can spend a few hours in the commode with *Uncle John's Bathroom Reader* until quitting time.

Nine hours without a bathroom break stuck in a partial pressure suit at altitude is a recipe for biological misadventure. No amount of training could convince the human bladder that it doesn't need to be urgently emptied at some point over a 540-minute-long span.

The solution was to outfit their suits with what is known as an indwelling catheter, defined as "a tube threaded up the length of the penis inside the urethra and into the bladder." The tube drains off urine that enters the bladder and empties it into a bag. It was as unpleasant as it sounds. Painful to insert, awkward to use, it was a textbook case of the cure being worse than disease.

The military may have been relatively hapless against the force of a full bladder, but defecation was another matter entirely. Beginning in the 1950s, the Central Intelligence Agency (CIA) devised a diet that would help U-2 pilots avoid the need to poop or pass gas. The CIA's "U-2 Pilot Physical Maintenance Control Program" suggested:

> *Controlled feeding of a high protein, low residue diet for mission pilots should begin twenty-four hours prior to takeoff. The objective of this controlled diet is to provide foods which can be almost completely absorbed from the gastrointestinal tract, thereby leaving a minimum of residue for the formation of feces and intestinal gases.*

Among the endorsed foods were coffee, tea, rice, eggs, soups, sweets, and cottage cheese. Those frowned upon included fried foods, fatty foods, cheese, milk, bread, pies, and pastries. What was difficult for the military pilots would prove to be grueling for NASA astronauts.

After Shepard's pissing contest, NASA hired space suit expert James McBarron to direct B. F. Goodrich—the company that designed the Mercury space suits—to incorporate a urinary-collection device. Goodrich delivered the Urine Collection and Transfer Assembly, or UCTA, in time for John Glenn's 1962 orbital flight. Astronauts donned the UCTA over the liquid cooling garment of their space suits, then hitched it to themselves using a roll-on cuff attached to a collection bag.

The user could empty the UCTA afterward through a one-way valve into a collection tank.

While rocketry was advancing by leaps and bounds—there was Project Orion, paralleling the Apollo effort, to build a starship powered by nukes—waste collection on the Apollo-era missions made an outhouse feel like a Kohler showroom. When Gordon Cooper flew the last Project Mercury mission in 1963, his capsule's systems started failing fast, one after the other. That forced him to take the wheel and manually steer Faith 7 through a perilous reentry into the atmosphere. Cooper landed safely in the Pacific Ocean, a few miles from his recovery ship, the USS *Kearsarge*. As successful as the flight was, the last-minute glitches cast a worrisome cloud over the mission. The problem, it turned out, was that his urine bag leaked. And some of the driblets seeped into the capsule's electronics. By the time of the Apollo crews, astronauts wore a somewhat similar condom-like pouch that hooked up to a hose. With a flick of a valve, they jettisoned the fluid into the vacuum of space. If that sounds weird, it is.

While Neil Armstrong was the first human to walk on the moon, Buzz Aldrin was the first to urinate on it. When he made a longer-than-expected jump from the lunar module, the Eagle, to the moon's jagged surface, his urine collector tore open on impact. Later on, when he relieved himself, the discharge spilled from the burst bag and filled one of his boots. Unlike Shepard, he made no recorded comment for posterity about his feat.

Handling urine was the easy part. Dealing with defecation plucked the last remaining sequins off the astronauts' glamorous image. While there were tweaks and adjustments made to it, NASA's "fecal containment system" didn't change much through the space race years.

On the Apollo missions, for example, an astronaut who had to defecate politely floated his way into a corner. Meanwhile, his companions would inch as far away as they physically could, which, in an Apollo command/service module capsule that measured 36.2 feet high with a 12.8-foot diameter, wasn't far.

Next, he often removed his clothes completely, shucking the entire cumbersome space suit. To collect his excreta, an astronaut was outfitted with a plastic bag in the shape of a tube. Around the bag was an adhesive, which he used to stick to his buttocks. Once he managed to get the bag to stay in position, he relieved himself, after which he could use an included "finger tube" to manually disentangle any waste that didn't make it all the way into the bag. Then he wiped himself with a tissue and a wet wipe, dumping those items into the fecal containment bag as well. When he was finally done—the process sometimes consumed forty-five minutes to an hour—he added the contents of an attached germicide pouch by squeezing the little sack from the outside of the bag to release its contents. Once the bag was sealed, the astronauts were stuck with it. There was no option for tossing it into space. Instead, the bags sat in the capsule until the mission returned to Earth.

Neither the stripping, nor the storage, were the nastiest parts, as the crew of Apollo 10—Tom Stafford, Gene Cernan, and John Young—discovered up close and personally on their way home from circling the moon in the May, 1969, "dress rehearsal" for Apollo 11.

Gene Cernan: Where did that come from?

Tom Stafford: Give me a napkin, quick. There's a turd floating through the air.

John Young: I didn't do it. It ain't one of mine.

Cernan: I don't think it's one of mine.

Stafford: Mine was a little more sticky than that. Throw that away.

Young: God almighty.

All three: (Laughter)

Often NASA is credited for its spin-off technologies that reshape the world around us; think running shoes and firefighting equipment, to name a few. But when it comes to space diapers, it worked the other way around. Inspired by Kimberly-Clark Corp.'s 1983 launch of Depends, which highlighted and captured a market of incontinent oldsters who still needed to remain active, NASA developed a Maximum Absorbency

Garment, or MAG, for the space shuttle program. Each MAG came with multiple thin layers of material that would first whisk urine from the body, after which sodium polyacrylate, a polymer able to absorb four hundred times its weight in water, locked the moisture away. Each MAG could handle 22.1 quarts of liquid in it, or about twenty-six times John Glenn's mammoth void in 1962. The shuttle came standard with a toilet, but MAGs would be needed for launch, reentry, spacewalks, and the occasional emergency—such as when the latrine on STS-1, the first shuttle mission, broke down, resulting in "clouds" of icebound fecal material drifting through the spacecraft.

While much improved over the crude facilities available to Apollo-era spacefarers, the bathroom on the ISS is an experience more akin to roughing it in the woods. When the ISS occupants use their $19 million, Russian-built bathrooms to urinate, a funnel/fan combination siphons their waste away into a yellow cone-shaped container. After about eight days of processing, the urine is transformed into drinking water for the space station.

Defecation is another matter altogether. The excrement drops down through a dinner-plate-sized hole on top of a silver can, where a vacuum fan speeds the waste away into a plastic bag. Once the bag begins to bloat with the combined discharge of the crew, it requires the astronauts to don rubber gloves and pack it down to a more-manageable form. Eventually the bag is packed on one of the cargo ships resupplying the ISS, where it's sent to a fiery demise in the Earth's atmosphere. Plumbing may be boring, but without it, no space exploration, from low orbit to the Sombrero Galaxy, will be possible.

CHAPTER ELEVEN

Covering Up: Space Blankets

Few people know that the ancients Greeks are as responsible as NASA for the everyday staple we now know as the "space blanket." Because if it wasn't for that first 26-mile-and-385-yard sprint run by the Greek soldier Pheidippides, a messenger dispatched from the Battle of Marathon to Athens to report the Hellenes' victory over the invading Persians, we might not have the silvery-gold sheets we universally recognize and use today, for everything from warming marathoners to shielding astronauts. That, and the conspiracy theory that later grew up around it, rivaling even the supposed collusion between director Stanley Kubrick and NASA to fake the Apollo 11 moon landing.

The material for what we now generically call the space blanket was registered as a trademark in 1952 by E. I. du Pont de Nemours and Company (DuPont). Called Mylar, it was basically plastic sheeting vapor-coated with aluminum particles. When the plastic was bonded to those particles, it formed a metallic material that reflected 98 percent of the heat and light that hit it. Not only was its reflectivity superb, but also its tensile strength was superlative: The individual molecules locked into place so tightly, Mylar was nearly impossible to tear, or even stretch. Moreover, Mylar was invulnerable to chemical reaction and easily formed an impassable barrier for moisture and air. It laughed off most solvents and acids. In short, Mylar was to industrial chemistry as Superman was to comic books.

That was the era when Big Science itself was as cool as Beyoncé and as pervasive as the Kardashians. Mylar was so revolutionary that in 1955

DuPont produced an industrial movie about the wonder material that has since joined the ranks of *Admiral Cigarette*, *The Stenographer's Friend: or, What Was Accomplished by an Edison Business Phonograph*, and other sponsored films that exalted business. In *Mylar: What's It to You?* the movie's title asks the audience in salmon-colored letters within the first minute, after which the friendly host swings a baseball bat at the viewer, stopped from breaking the fourth wall—and the viewer's head—wide open by an invisible shield of Mylar that he marvels is "as thin as a human hair."

The film mixes two familiar outposts of the Eisenhower-era pop-culture landscape then: the variety-act format of *The Ed Sullivan Show* with the torture tests of Timex's John Cameron Swayze commercials. First a man dressed in a gaucho/circus outfit bounces on a trampoline made of a sheet of Mylar, stretched thin to the point of transparency, giving it what the narrator admiringly calls "a pretty rugged beating." Following that, a bowler hurls a bowling ball into a wall of Mylar "at full speed!," with no effect—even after the barrier's been frozen supposedly to the point of brittleness with blasts of carbon dioxide from a fire extinguisher. A sultry Marilyn Monroe–esque acrobat twists on a trapeze supported by thin strands of the miracle material, after which the platonic suburban ideal of a 1950s housewife scrubs a kitchen wall made of it, then segues into a gown of the space-age material for a night out. Decorative women in maroon berets and hot pants hold up sculpted forms of Mylar, while the host reminds all the little ladies in the audience that with this miracle material, "Draperies can take on exciting new effects!"

Mylar was first used commercially in the 1950s as an unconventional wall covering in bathrooms, as well as in the production of magnetic audio- and videotapes, and even drumheads. In 1955, the Eastman Kodak Co. wove Mylar into a photographic film called Estar, whose physical durability allowed the high-altitude U-2s to use it for their reconnaissance flights.

Still, Mylar languished for a few years in a kind of conceptual ghetto, the way earthshaking inventions often do at first—barely making a ripple when they're fresh out of the box, then becoming as ubiquitous as water, air, or cell phones. Western Union was a paradigm of that industrial myopia, spurning an offer from Alexander Graham Bell to buy

Echo 1 satellite inflation test. NASA

the telephone's patent for $100,000, sneering that it was "hardly more than a toy." A British parliamentary committee shared that shortsightedness, scoffing that Thomas Edison's lightbulb was "good enough for our transatlantic friends . . . but unworthy of the attention of practical or scientific men."

Unlike telegraph companies or John Bull, however, NASA quickly saw the benefits of a material that could at once repel the concentrated heat of the sun and the subzero cold of outer space. In 1960 the agency put the material to work when it launched the first communication satellite, Echo 1. Measuring 100 feet in diameter and weighing 146 pounds at launch, the Echo 1 was essentially nothing more than a giant Mylar balloon. Once in orbit, the satellite acted as a passive reflector that could bounce telephone, radio, microwave, and television signals from one ground station to another. Its reflective Mylar skin shined so bright,

Echo 1 was visible to the naked eye over most of the Earth, flaring like a birthday candle.

Supplying NASA with Mylar was the National Metallizing Division of Standard Packaging Corp., headquartered in Cranbury, New Jersey. Like the company, NASA realized the material possessed an all-purpose, duct-tape-like adaptability to a variety of purposes. (Later, Standard Packaging would market Mylar in more-mainstream pursuits, including building insulation.) Perhaps none were as live-or-die as space suits, which had to protect astronauts' fragile flesh from the inhospitable extremes of radiation and temperature. While the space suits for the Mercury program retained the designs of pressurized flight suits typically used in high-altitude aircraft, their successors for the Apollo program were an altogether different breed. Understanding that the demands of the moon landing would be far more rigorous, NASA constructed Apollo suits from a twenty-one-layer combination of twelve synthetic materials, among them Kevlar, Nomex, Kapton, and, of course, Mylar, endowing the ensemble with more strength and greater tolerance toward life-threatening temperatures, which on the moon swing pendulum-like between 233 degrees below zero and 253 above it.

Buried in our communal memory, however, is a spec for the suit that was grounded in tragedy. Following Roger Chaffee, Ed White, and Gus Grissom dying from the toxic gases released when a fire broke out in their Apollo 1 spacecraft during a ground test in Cape Kennedy's Launch Complex 34 in 1967, NASA dictated that the suits had to weather temperatures exceeding 1,000 degrees Fahrenheit.

Mylar further ensured the safety of the astronauts in their space suits with a coating over the base of the Apollo lunar-landing vehicles. Though moon voyages sputtered to a halt after 1970's Apollo 17, Mylar's fortunes continued to soar. By the late 1960s, Mylar and cellophane produced nearly two-thirds of DuPont's Film Department sales. By the 1970s, Mylar had become the company's best-selling film. Its terrestrial successes were matched by its dramatic off-world achievements: On May 14, when the Skylab space station was launched into orbit, it somehow lost its micrometeoroid shield. When the protective barrier tore off, it also took a main solar panel array with it. Losing those created a

Another view of the Echo 1 satellite during an inflation test. NASA

dire-verging-on-disastrous ripple effect, because not only was the electrical power crippled, but the lack of shielding also allowed unimpeded sunlight to heat the air temperature inside the station, causing it to climb toward +130 degrees Fahrenheit, or the internal temperature of beef roast cooked medium. Within a short time, the food, and other equipment inside, would deteriorate. And the high-flying platform's materials might even begin to spew toxic gases.

On May 25, 1973, a crew of three astronauts launched toward Skylab to erect an emergency substitute protective shield for the one that had snapped off during the launch of the short-lived space station's ascension into orbit. The custom-made replacement consisted of layers of nylon topped by a stratum of Mylar.

Mylar has wound its way into everyday life, from hydroponic gardening to sailboat racing. Now a product of a joint venture, DuPont Teijin Films, Mylar is a component of space suits, heat shields, and solar sails. As Mike Weiss, the technical deputy program manager for the Hubble telescope, once put it, "Thermal blankets are to spacecraft as clothes are to people."

It's also baked into products ranging from the backing on solar panels to the aluminized suits of firefighters. But none of its varied uses granted it immortality like the race an exhausted Greek soldier first made 2,600 years ago.

In the 1970s the Association of International Marathons and Road Races began distributing space blankets manufactured by Metallized Products at a number of races to help contenders ward off hypothermia. One of the races included the 1979 New York City Marathon, where David Deigan—a former employee of National Metallizing—took part. In a textbook case of "what goes around comes around," Deigan saw dozens of runners cocooned in the shimmering blankets after he crossed the finish line, and realized that here was a marketing opportunity whose potential was as high as Skylab or Echo. Incorporating his company AFMInc (Advanced Flexible Materials) in 1982, he began producing the Mylar blankets under the name Heatsheets, selling 700,000 of them annually, now as de rigueur in races as compression socks and water bottles.

Even more profoundly, space blankets are as much a part of the refugee experience as tents and thousand-yard-stares. So deeply entwined are they with the imagery of the world's 68.5 million refugees, they serve as a metaphor for the exiles' plight: Italian artist Cristina Ghinassi used them in a performance-art piece in Istanbul, while artist Ai Weiwei encouraged actor Charlize Theron, Pussy Riot member Nadya Tolokonnikova, and various other celebrities/1-percenters attending his Cinema for Peace fund-raiser in Berlin to don the blankets en masse for a group photo op.

The story of space blankets doesn't end with brave new worlds conquered and the wretched of the Earth sheltered. Instead, it trails off in a cultural cul-de-sac stocked with enough paranoia for a Dan Brown novel. Indeed, the name this phenomenon generally goes by—Black Knight—would relieve Brown and his publisher of having to think up a marketable title. Depending on the source, the story begins about 13,000 years ago, when an alien civilization shot a satellite—the aforementioned Black Knight, whose name is as cool as it is of hazy origin—into Earth orbit, while others claim it started in 1923, when Nikola Tesla said he received a Chariots-of-the-Gods communiqué through a wireless receiver of his own design. (Tesla thought Martians were at the other end of his cosmic phone, which some hold was actually the orbiting Black Knight contacting the inventor.) In the 1960s, others pointed to an anomalous satellite in polar orbit—most were in equatorial orbit then—as proof that the Black Knight hovered over us, an ancient eye in the sky monitoring us since the days when we first started domesticating animals. Declassified documents later identified it as a Cold War satellite monitoring the Soviets.

In either case Black Knight picked up steam after Col. Jerry Ross and Dr. James Newman's 1998 spacewalk outside the International Space Station (ISS). While the duo were attempting to install thermal blankets as protective covers on the ISS's four trunnion pins, to inhibit heat leaks through the pins' exposed metal, one slipped off Colonel Ross's space suit, tumbling into the void. After alerting Mission Control to the runaway blanket, a shuttle crew member took video of the escaping object.

Like it does with all space debris above a certain size, NASA even issued it an official object number: 025570. True believers perceived that the purported visual record of the fleeing blanket was not so much an accident but an Aliengate cover-up. After all, photos of the "blanket" suggest the Klingon idea of a shark, designed by extraterrestrial minds and fashioned by alien hands unfamiliar with Euclidian geometry. Nothing so plain as a covering that sheaths humanity in space, and safeguards it here on Earth.

CHAPTER TWELVE

Cold Calling: ETs

WHERE ARE THEY? THE BUG-EYED MONSTERS, THE LITTLE GREEN MEN, the grays, and the dust-up at Roswell notwithstanding, shouldn't aliens be here by now? Didn't John Glenn supposedly see . . . something from the Friendship 7, his capsule on the Mercury-Atlas 6 flight, in 1962? Didn't Buzz Aldrin, en route to the moon on Apollo 11, notice a light out the window that appeared to be shadowing the craft's path?

Drs. Philip Morrison and Giuseppe Cocconi suggested using microwave radio for otherworldly communication in their magazine article "Searching for Interstellar Communications" in the September 1959 issue of *Nature*. The search for extraterrestrial life took on the boundless optimism of the search for El Dorado or the Holy Grail.

At nearly the same time, Harvard University astronomer Harlow Shapley hazarded a rough guess on the number of inhabited planets in our entire thirteen-billion-year-old universe.

The universe has 10 million, million, million suns (10 followed by 18 zeros) similar to our own. One in a million has planets around it. Only one in a million, million has the right combination of chemicals, temperature, water, days and nights to support planetary life as we know it. This calculation arrives at the estimated figure of 100 million worlds where life has been forged by evolution.

Less than a year after Morrison and Giuseppe Cocconi's *Nature* article, Dr. Frank Drake at the National Radio Astronomy Observatory in Green Bank, West Virginia, initiated Project Ozma, the first SETI (search for extraterrestrial intelligence) quest, with an 85-foot antenna tuned to the 21-centimeter, 1,420-megahertz (MHz) line of neutral hydrogen. This is a particular wavelength of radio emission by neutral hydrogen, the most common element in the universe. Drake, alongside Cocconi and Morrison, reasoned that other technological civilizations would likely monitor it as well to suss out the telltale signs of intergalactic neighbors. Bernard Oliver, a vice president of research and development at Hewlett Packard Corp., in 1971 dubbed the 21-centimeter line a "watering hole"—meaning, a common place to meet and talk, like a body of liquid refreshment in the Serengeti around which other creatures would naturally gather.

Drake and the others too reasoned that other intelligences might see the 21-centimeter line as a logical landmark in the radio spectrum. He perused frequencies near the wavelength for nearly four months, but all he heard was the sound of silence.

His criteria for what would constitute an artificially produced signal by an intelligent mind had to hit several benchmarks before it could be considered a text message from the stars:

1. The number of particles received must significantly exceed the natural background count.

2. The signal must exhibit some property not found in natural radiations.

3. In addition, the radiation should require the least radiated power.

4. It should not be absorbed by the interstellar medium or planetary atmospheres.

5. It should not be deflected by galactic fields.

6. It should be readily collected over a large area.

7. It should permit efficient generation and detection.

8. It should travel at high speed.

9. It should normally be radiated by technological civilization.

In 1961, Drake and the National Radio Astronomy Observatory hosted the first SETI Conference in Green Bank. There, Drake introduced his eponymous Drake Equation for guesstimating the number of advanced technological civilizations spread out in the Milky Way Galaxy's 400 billion or so stars:

$$N = R^* \cdot fp \cdot ne \cdot fl \cdot fi \cdot fc \cdot L$$

where:

N = the number of civilizations in our galaxy with which communication might be possible (i.e., which are on our current past light cone);

and

R* = the average rate of star formation in our galaxy

fp = the fraction of those stars that have planets

ne = the average number of planets that can potentially support life per star that has planets

fl = the fraction of planets that could support life that actually develop life at some point

fi = the fraction of planets with life that actually go on to develop intelligent life (civilizations)

fc = the fraction of civilizations that develop a technology that releases detectable signs of their existence into space

L = the length of time for which such civilizations release detectable signals into space

His contentious statistical method was meant less to provide a hard and fast number as it was to spark a debate about how many civilizations might be out there, whether they be Tribbles or Time Lords. The only reference point we had for "radio loud" culture was our own. At this point we had been broadcasting for well more than a century: The first radio program broadcast was on Christmas Eve, 1906, when Reginald Fessenden of Ocean Bluff-Brant Rock, Massachusetts, played "O Holy Night" on the violin and read a passage from the Bible to ships sailing the Atlantic. Hopes were high at the conference that we might be just one more plant in the botanical garden of technologically evolved species.

People never seem as optimistic as when they attempt the improbable, as if the energy of their hopes will translate into a tangible ROI. "At this very minute, with almost absolute certainty, radio waves sent forth by other intelligent civilizations are falling on the earth," Drake wrote in *Intelligent Life in Space*. "A telescope can be built that, pointed in the right place, and tuned to the right frequency, could discover these waves. Someday, from somewhere out among the stars, will come the answers to many of the oldest, most important, and most exciting questions mankind has asked."

The members of Project Cyclops, the name given to a 1971 NASA study team, wanted to ask those questions. They would have spearheaded the design and construction of an array of up to one thousand radio telescopes to detect Earth-type radio signals up to one thousand light-years away, but the venture's price tag—$6 to $10 billion—doomed it before it could leap from paper concept to physical reality.

Six years after the proposed Project Cyclops, the Ohio State University Radio Observatory telescope—its nickname "Big Ear" bestowed for its size of three football fields—detected the famous "Wow!" narrowband signal from the constellation Sagittarius. Astronomer Jerry Ehman discovered the incongruity a few days later after Big Ear recorded it, while routinely poring over the logged data. So intense was the full seventy-two-second-long signal—it was thirty times stronger than the cosmic background noise, and right at the 1,420 MHz hydrogen line frequency—Ehman drew a circle around it and wrote the now-legendary descriptor "Wow!" on the computer printout. Arguments have long raged and simmered and boiled up again over its meaning (if there is any). Some now think it was produced by a spy satellite or a comet, but the mystery endures.

Eventually it became apparent that the universe talks to us no more than a skeleton could. Perhaps the universe's cool kids were not listening on the supposedly popular 21-centimeter / 1,420 MHz hydrogen line after all.

With interest in the space race waning unexpectedly after the moon landing, NASA's budget slumped to about $3.8 billion in 1970, from a high of $5.9 billion in 1966. Even that amount, representing about a 35 percent drop, wasn't enough to shield the agency's budget from calls to use their money on an endangered Earth rather than a lifeless moon.

With the near disaster of Apollo 13 further dampening the interest in moon missions even more, and NASA canceling the program after Apollo 17, it shouldn't be surprising that the search for neighbors in space become a casualty as well. Senator Proxmire of Wisconsin bestowed NASA with his Golden Fleece Award for governmental waste in 1978, for its $14 million proposal to scan the skies for signs of alien life. In 1993 NASA had to shutter its SETI-focused High Resolution Microwave Survey (HRMS) after just one year of operations, over Congress's concern that the ten-year program would be a waste of its $100 million budget.

Two years after NASA pulled the plug on HRMS, Harvard University started using its 84-foot steerable radio telescope for Project BETA (billion-channel extraterrestrial assay). After the telescope was damaged in a 1999 windstorm, the effort shut down for good. Two years before BETA stopped, Ohio State's Big Ear SETI project, which *The Guinness Book of World Records* formally recognized as the longest search for extraterrestrial intelligence in history, was torn down to make way for a golf course.

Science has told us that intelligent life in the universe is nowhere yet to be found: Venus was worse than Hell, because at least Hell didn't come with clouds of sulfuric acid. Mars was an emaciated husk with the largest dust storms in the solar system. COROT-7B, 489 light-years from Earth, has temperatures of +4,580 degrees Fahrenheit, and it rains rocks there. HD 189773B is closer, a run-to-the-store 63 light-years away, but has torrents of glass grains that fly around in the planet's 4,000-mph wind.

So maybe we were getting it all wrong. Perhaps the universe is a cemetery, its planets headstones, its galaxies mausoleums. Maybe, according to the contrarian thinkers, instead of listening for extraterrestrial conversations, we should be looking for their parking lots and skyscrapers. After all, couldn't edifices built by human hands, even without the benefit of automation, supposedly be seen from the moon?

Reality check: You're as likely to be able to see the Great Wall from the moon's surface as you would a human hair from 2 miles away. Even from a low Earth orbit of 160 to 350 miles, you can't see it, because the Great Wall's colors blend in too well with the natural topography around the Wall. But other massive objects—the Great Pyramids at Giza, parts of the Kennedy Space Center, the 64,000-acre-sprawl of highly reflective plastic greenhouses in Almeria, Spain—can sometimes be discerned from those considerable heights. NASA's Lunar Reconnaissance Orbiter—currently circling the moon—can pick out Neil Armstrong's boot prints from his 1969 jaunt on its surface. Why wouldn't aliens, some of whom might be capable of faster-than-light travel, build something so gargantuan we could see it, even from distances measured in parsecs? As the author and historian of science Michael Shermer put it in his riff on a famous Arthur C. Clarke quote, "Any sufficiently advanced extraterrestrial intelligence is indistinguishable from God."

Maybe "God" is more like an architect with a rich client. More than fifty years ago, in 1964, Soviet astronomer Nikolai Kardashev became intrigued by this possibility when developing the eponymous Kardashev scale. In it, he lumped civilizations into three main types—Types I, II, and the big kahuna, III.

- A Type I civilization—also called a *planetary civilization*—can use and store all of the energy which reaches its planet from its parent star.

- A Type II civilization—also called a *stellar civilization*—can harness the total energy of its planet's parent star. This would mean a mega-massive construct that surrounds the entire star, which then transfers its energy to the planet(s).

- A Type III civilization—also called a *galactic civilization*—can master all energy on the scale of its entire host galaxy.

Based on the above, whatever we would be looking for, it was likely going to be big. The best way to get a grip on the concept is to look at the marvelous examples science fiction is rich with. In that genre of starfaring storytelling, it's easy to find examples of alien engineering that make the pyramids look like a pillow fort built by bored children on a snow day. The best, or at least the best known, comes from *2001: A Space Odyssey*. In that classic of boundary-stretching cinema, the mute and millions-of-years-old monolith plays a God-like Henry Higgins, whose Eliza Doolittle—humanity—turns out to be something of a disappointing protégé. Consequently, in the movie and book sequels—*2010: The Year We Made Contact* and *2010: Odyssey II*, respectively—the unfathomable aliens turn Jupiter into our solar system's second sun, and terraform its fourth-largest moon in a kind of *Europa Extreme Makeover: Evolutionary Edition*. In the recent *The Expanse* television and book series, aliens two billon years gone launch a moon-sized probe that becomes stuck in Saturn's orbit. The satellite harbors a protomolecule that can shape planets like Play-Doh and rip holes in the tensioned fabric of space. Still, those examples exist on the cusp of the unbelievable, or at least the improbable. It remains something more likely to be seen on the Syfy channel than the NASA website.

Or maybe not. Freeman Dyson—who also worked on a nuclear-powered spaceship—first explored this idea in 1960. Dyson's published two-page paper in the journal *Science* in 1960 was titled "Search for Artificial Stellar Sources of Infrared Radiation." Dyson proposed probing for a sun-sized solar power collector, similar to what Kardashev would essay a few years later with his Type II civilization. The Dyson sphere, as it's now known, is popularly thought of today as a solid covering over a star, sucking out its juice for whatever purposes the aliens intend. Dyson himself thought the idea of a solid sphere impractical. Instead he envisioned it as a cloud of objects encircling a star and drawing its energy for the benefit of its makers. Still, the more ambitious, if erroneous, concept persists. The closest we've come to Dyson's grand idea is an episode of *Star Trek: The Next Generation*, called "Relics," where the USS *Enterprise* encounters a solid Dyson sphere with a diameter of 200 million kilometers, and an internal surface area totaling about 250 million Earth-like planets.

Still, we haven't stopped searching. About ten years ago Richard Carrigan at Fermilab in Illinois published his hunt for detected spheres around sun-like stars only within one thousand light-years of Earth. More recently, Jason Wright at Pennsylvania State University sifted through data from two space-based infrared observatories, some 100,000 nearby large galaxies in the Wide-Field Infrared Survey Explorer and the Spitzer space telescope, to hunt for signs of the waste heat that Dyson spheres would theoretically leave behind.

In 2018 scientists began kicking around the idea of using the European Space Agency's Gaia mission to seek Dyson spheres. The Gaia spacecraft, which have surveyed 1.7 billion objects in a 3-D color map of the cosmos, might be capable of detecting a smaller Dyson sphere out to around one hundred light-years, but larger ones virtually anywhere. So far, nothing.

Sadly, the Apollo years also left us empty-handed. There were no aliens on the moon. The fireflies John Glenn saw from Friendship 7 were first thought to be ice crystals from the hydrogen peroxide control rockets fired. Other astronauts on subsequent flights witnessed similar "fireflies," which turned out to be clouds of debris that gathered around the capsules, reflecting sunlight and creating the effect of sparkling space pixies. Buzz Aldrin's UFO was nothing more, he later admitted, than either the rocket the Apollo 11 crew had separated from, or part of the four panels that moved away when they extracted the lunar lander from the rocket.

Maybe with our own attempts at launching objects, aliens might take notice. So far the best we've been able to do is Elon Musk launching one of his electric cars into space. In February of 2018 Musk's SpaceX company put a cherry-red Tesla Roadster on board its Falcon Heavy rocket's inaugural launch. The Roadster—with a dummy in the driver's seat called Starman (a nod in the direction of the David Bowie song of the same name)—has now reached beyond the orbit of Mars. The Roadster and Starman will come near Earth in 2091, and will crash into either Venus or Earth itself a few tens of millions of years from now.

Still, hope springs. A few years ago a patent was filed for a "planet kidnapping system." The inventor explains (in what is probably the tone of voice hostage negotiators hear a lot of) that it employs lasers to push asteroids to hit a given planet of moderate size, so that the users can change the target planet's orbit and move it to another location, like shoving Mars out to Pluto's trajectory. (That would have made an interesting sequel to *The War of the Worlds*.) When it comes to contacting aliens with engineering too big to ignore, lunacy is just brilliance taking its time.

CHAPTER THIRTEEN

Fueling Around:
Land-Mine Deactivation

ONLY 24 PEOPLE HAVE EVER FLOWN TO THE MOON, AND JUST HALF HAVE walked what Buzz Aldrin called its "magnificent desolation." Roughly 560 individuals have vaulted into space, a privileged roster including American businessman Dennis Tito who in April of 2001, at the age of sixty, became the first space tourist, spending nearly eight days in space, much of it aboard the International Space Station, for a sum of $20 million. In 2008, Richard Garriott, a video-game designer and an investor in Space Adventures Ltd. (whose father, Owen Garriott, curiously enough, was an astronaut), spent twelve days in space, for $30 million, consuming almost the entirety of his fortune. In 2018, Elon Musk announced that his Space Exploration Technologies Corp. will launch Japanese clothing-company billionaire Yusaku Maezawa (who once blew $110.5 million on Jean Michel-Basquiat's 1982 painting, *Untitled*) into the starry void.

Compare those sparse figures to the number of those who have scaled Mount Everest—nearly 5,000—which makes ascending the 29,029-foot-high Himalayan elevation look like a club that lets just about anyone in.

"The reality is the majority of us will not get off this planet," astronaut, engineer, and *Star Trek: The Next Generation* actor Mae Jemison once said. "So the long run is, some kind of space exploration has to benefit us here on Earth." Something that offers more advantage to the survival of the species than a cordless drill. Something that makes life just a little less nasty, brutish, and short for the 99 percent. Something, for example, that neutralizes land mines.

According to the *Landmine Monitor 2017*, these titular explosive devices caused 8,605 casualties that year, among which at least 2,089 people died. The year before that, in 2016, the victim count spiked at 9,228, the most casualties recorded since 1999. Included in that morbid accounting was the highest number of child victims ever documented. Was this the sort of societal problem NASA could help with? Or was it just another reason that NASA's efforts might be better focused on starting over on another world, while this one winds down in a Mobius strip of violence and malice on an industrial scale?

Originally developed by North Carolinian Gabriel Rains, land mines were first wielded by Rains during the Second Seminole War in Florida in 1840. Later Rains, once a student at West Point, gifted in chemistry and prodigious in engineering, switched sides from the Union to the Confederacy, where he put his invention to yet more murderous use. During the 1862 retreat from the Union's siege of Yorktown, Virginia, Confederate soldiers planted the invention, sometimes called the "Rains Patent," to the surprise and horror of their pursuers, a battlefield reality that would become familiar to soldiers as distant in time as the Argonne and Afghanistan. Sometimes known as "torpedoes" or "subterranean shells," the weapon evolved at the super speed of all revolutionary armaments. A military definition of land mines is, essentially, an explosive device designed to annihilate or impair equipment or humans. Equipment generally refers to boats, aircraft, and ground vehicles. Mines are detonated by one of three means—by remote control, by the passage of time, or by the physical movement of the intended target. Mines can basically be divided into two self-explanatory types: antitank and antipersonnel.

Once they entered the mainstream version of combat, land mines were initially used as a defensive weapon, shielding areas such as important borders or strategically significant bridges to hector the advances of enemy forces. Later, they progressed into the platonic ideal of pure terror that found their apogee of function was in maiming rather than killing— because more resources are consumed caring for the injured than the dead. Land mines are in essence the perfect sleeper agents—never slacking, always vigilant, ever lethal. Of an estimated 100 million land mines still active in sixty-four countries, Egypt is saddled with an astonishing 23 million; followed by Iran, with 16 million; Angola, with 9 to 15 million; Afghanistan, with 10 million; and Iraq, also with 10 million. Worldwide, there are perhaps 300,000 children alive who have been severely disabled by land mines. Currently, there are about 2 million more planted each year, with 100,000 mines removed annually worldwide. It is much easier to plant mines than it is to eradicate them: Mines cost between $3 and $30 to plant, but $300 to $1,000 apiece to remove.

Getting rid of mines has proven to be as difficult as getting rid of another mass murderer, the mosquito. Efforts by the International Campaign to Ban Landmines culminated in the 1997 Convention on the Prohibition of the Use, Stockpiling, Production, and Transfer of Anti-Personnel Mines and on their Destruction, which was passed in 1999. Also known as the Ottawa Convention or the Mine Ban Treaty, it bound its signees to eliminating antipersonnel devices, while also prohibiting the use, production, stockpiling, and transfer of land mines. More than twenty years on, it has yet to be universally adopted. Even though more than 162 nations have signed it, 33 others, including the United States, Russia, China, India, Myanmar, Cuba, Egypt, India, Israel, and Iran, have remained holdouts. The US refusal to sign the treaty boils down to erratic North Korea. If North Korean forces invade South Korea, the United States wants to be able to deter an invasion with the weapons of its choice.

With Islamic State (ISIS) militants in Syria, Iraq, and Afghanistan scattering these devices like violent seeds, the problem is far from even plateauing. Finding and neutralizing land mines is an effort as slow as it is dangerous and seemingly insurmountable. In 2014, when Cambodia cleared roughly 20,000 land mines, covering 21 square miles of land, that effort accounted for 27 percent of the worldwide total clearance for that year. By that reckoning, and if the number of land mines was frozen at 100 million, it would still take more than twelve centuries to clear them all.

This is where NASA inventiveness comes in. In 1999, the same year the Mine Ban Treaty went into effect, NASA offered, if not a solution to the devastating problem of land mines exactly, then at least a flicker of hope that the threat they represent could be reduced. And all because of a logistical need to overcompensate.

In making sure that its space shuttle missions had sufficient fuel on hand for each flight, NASA allocated a given percentage of extra propellant to be available. But once it's prepared, the propellant congeals quickly. When it's in its hardened form, it cannot be saved for use in another launch, and so it goes to waste.

This didn't mean it was actually wasted, however. While the fuel's thickened form might mean it was useless for the space shuttle, it was ideal for a secondary purpose few had ever anticipated: detonating land mines from a safe perch, far away.

It was a marvelous idea, seemingly one that could do no wrong, and much good. But a technology doesn't always get used unless the time and moment in history pave the way for it. After all, it took thirty years for electricity to reach 10 percent of the country, and it wasn't until the 1960s that landline telephones—invented in 1876—reached 80 percent of US households.

The NASA technology might have remained moot and never found its niche if not for Princess Diana. Putting the lie to John Osborne's famous snipe that "Royalty is the gold filling in a mouthful of decay," with her work with AIDS patients the 1980s and early '90s preceding her, Diana was about to engage in the work that would define her legacy.

In 1997 the Princess of Wales visited the Republic of Angola, which reportedly had the highest percentage of amputees anywhere in the world, mostly due to land mines planted during the generation-long civil war. Wearing chinos and body armor and walking dangerously through an active minefield, Princess Diana, more than any single person, ripped the veil of legitimacy off land mines. "Her actions," said a founder of the Mines Advisory Group, "spurred what was perhaps the fastest-signed arms-control agreement in history, the Mine Ban Treaty."

Diana had unknowingly cleared a path for NASA. Through a memorandum of agreement between NASA's Marshall Space Flight Center and the propellant maker, Thiokol, the manufacturer received the green light to use the scrap fuel, more formally known as reusable solid rocket motor propellant. Even before the first mine was deactivated, NASA was realizing actual benefits. With this process, the agency no longer had to deal with the problem of trying to dispose of the toxic substance without polluting the environment.

The result was the Demining Device, developed by Thiokol in partnership with DE Technologies Inc., which neutralized land mines in a way that was safer and more efficient. Before then, mines could be disarmed with two approaches. The first was to set off the mine directly by direct physical contact of some kind. This forced people to come into close proximity to the mines, thereby risking their lives. The second approach was remote detonation, which involves using an explosive agent (C-4, say, or TNT) to blow the mine. But that solution added an explosion on top of the mine itself blowing up, resulting in more shrapnel and debris hurled even greater distances, injuring civilians or damaging property.

The Demining Device offered a third way. Triggered by a battery-powered electric match from afar, it generates a flame with an average temperature in excess of +3,500 degrees Fahrenheit, scorching a hole in the land mine's casing. The superheated flare simply burns out the explosive in the mine, leaving everyone safe. Even on the rare occasions when the mine unexpectedly blew, the de-mining personnel were distant enough to be out of harm's way. After testing the Demining Device on an assortment of antipersonnel and antitank land mines, the US Naval Explosive Ordnance Disposal Technology Division shipped several of them to Kosovo and to Jordan, two countries long under siege from mines' seemingly never-ending churn of casualties. Perhaps the single greatest reason behind its success was the brutal calculus of dollars and cents. Where it might formerly have cost up to $1,000 apiece to remove one mine the traditional way, the flare's price—around $6 apiece—made it an economical way to moderate the world's misery.

"NASA should start thinking about this planet," Wally Schirra once said. It did, and the proof is those whose bodies were never mangled and those whose lives were not lost.

CHAPTER FOURTEEN

Faster than a Speeding Bullet: Nuclear Rockets

THE ONLY THING SLOWER THAN EVOLUTION IS SPACE TRAVEL. IN REAL life, spacecraft move with the swiftness of snails traveling through a sea of peanut butter. Apollo 11 took slightly more than four days—102 hours, 45 minutes, and 40 seconds—to reach the moon, 240,000 miles away.

Things haven't changed much since then, or gotten much faster. Take the New Horizons mission to the sometimes-planet Pluto and the astronomical object 2014 MU69, both located in the Kuiper Belt. Launched by NASA on January 19, 2006, the interplanetary space probe accelerated quickly toward the demoted planet at a speedometer-busting 100,000 miles per hour. That seems pedal-to-the-metal fast, until you realize it took until July 14, 2015, for the spacecraft to reach the dwarf planetoid. It didn't even reach 2014 MU69, a billion miles past Pluto, until January of 2019.

(Side note: Fortunately, 2014 MU69 has the much cooler, crowd-sourced nickname of "Ultima Thule," a term from medieval times meaning "beyond the known world." NASA selected it from a pool of 34,000 names offered up by 115,000 participants, who both proposed and voted on the myriad names in a contest. The most popular, however, was "Mjolnir," aka Thor's hammer in Norse mythology, and the better-known and more-lucrative Marvel Comics universe. Also popular was "Tiramisu.")

The $700 million Earth-to-Pluto mission is typical of our current spacecraft, which in some ways hardly seems different from Robert Goddard's 1926 10-foot-tall rocket that soared to about 40 feet at a blazing speed of 60 mph. While a Saturn V rocket—the one that took us to the moon, with its approximately three million individual parts—may seem as different from Goddard's as a butterfly is from an F-16, their truest divergence may be only in the amount of chemical fuel they burn. With current technology, a trip to Mars, launched when the Earth and the red planet are at an optimal distance, could take a ship somewhere between six to nine months to complete.

This excruciatingly slow pace of rocketry continues to be a grave disappointment to those of us who ask: Where is Captain Kirk's USS *Enterprise* NCC-17901 that could tootle along at 512 times the speed of light? Or the Death Star, crushing rebels at a velocity of 1,142,500 times the speed of light? Or Mel Brooks's Eagle 5 from *Space Balls*, whose liquid Schwarz fuel powered it to run almost 180,000 times as fast as the aforementioned *Enterprise*? For all the Moore's Law advances in computer-related technology that seem to occur as often as we look at our phones, propulsion systems seem barely past the equine- and steam-based eras of achieving advanced momentum.

Others were also asking: Why can't we go faster? A lot faster. Unbeknownst to the world at large, early on in the race to the moon, several US government agencies, scared by the specter of Sputnik streaking over a Red sky with the white and blue subtracted from it, sponsored Project Orion—a craft powered by a nuclear thermal propulsion system.

Project Orion's beginning reaches back to the final days of the Manhattan Project, where Stanislaw Ulam and Frederic de Hoffmann were working on the effort to deliver the first atomic weapons. In 1944 Ulam, a mathematician, and de Hoffmann, a nuclear physicist, began tinkering with the idea of spacecraft driven by atomic bombs for space. The research and development of the initial weapons used in Hiroshima and Nagasaki, Little Boy and Fat Man, respectively, showed that the massive, almost fairy-tale-like explosive energy the bombs released might be tamed into precision-controlled thrust, pushing planes and rockets to unthinkable velocities. Chemical rockets were crude and wildly inefficient, needing about 16 tons of fuel to place 1 ton of payload into orbit. Nuclear thermal propulsion, though, was gauged at one hundred times more powerful than chemical systems of comparable weight.

This idea—probably as much a leap over standard rockets as the Wright Brothers' Flyer was over a Montgolfier Brothers' balloon—found more official form in a once-classified report Ulam had co-authored in 1955 called "On a Method of Propulsion of Projectiles by Means of External Nuclear Explosions." One key element in the article that would recur in several versions of the nuclear rocket was the concept of dropping a series of atomic bombs from its rear:

The bombs are ejected at something like one-second intervals from the base of the rocket and are detonated at a distance of some 50 meters from the base. Synchronized with this, disk-shaped masses of propellant are ejected in such a way that the rocket-propellant distance is about 10 meters at the instant the exploding bomb hits it.

The bombs would be trailed by disks made of a solid propellant. When the bombs detonated, they would boil away the hard disks and convert them into hot plasma. The expanding plasma would thrust against a pusher plate, safeguarding the vulnerable crew inside, and plunge the ship forward at ever-increasing speeds. But as the report quietly states, almost *sotto voce*, in another section, "The critical question about such a method concerns its ability to draw on the real reserves of nuclear power liberated at bomb temperatures without smashing or melting the vehicle."

The principle underlying the Orion concept can even be credited to nineteenth-century German inventor Hermann Ganswindt. In 1881— that is, a few years before Karl Benz invented the first gasoline-powered car—Ganswindt proposed a spacecraft that would be driven by tossing dynamite cartridges into a giant metal chamber. When the cartridges were detonated, half of their force would be expelled, while the other half would strike the top of the chamber to provide the reaction force moving the ship through space. Situated below the chamber was the crew's quarters, which would be spun constantly to create an artificial gravity. In a tip of the historical hat, the International Astronomical Union named a lunar impact crater near the southern pole of the moon's far side for Ganswindt.

Ulam and the others who toyed with the idea were nevertheless ahead of their time, and they weren't alone. In the years following World War II, the liberated atom seemed like it could usher in an engineered heaven of plenty and fortune. In a 1954 address to science writers, Atomic Energy Commission (ACC) chairman Lewis Strauss had implied nuclear energy would someday be "too cheap to meter." That phrase, which over the decades has become synonymous with overconfidence and naiveté, was perhaps closer to a daily reality than we imagine.

The idea of rockets with 10 billion horsepower that would have us touring Venus and sightseeing on Saturn spread like a Cambrian explosion across the scientific and military establishments. At first, many related studies, like those from North American Aviation and the Douglas Aircraft Co., focused on ICBMs with nuclear-powered engines for defense, but the enthusiasm collided with reality soon enough. Then-secret reports on the abovementioned companies in 1946 concluded that nuclear-powered ICBMs would require a reactor running at temperature of +5,700 degrees Fahrenheit—well beyond the capacity of contemporary materials to withstand.

And yet, nuclear-powered rockets were treated not just like an idea whose time had come, but also as an idea whose time was overdue. The ACC (later succeeded by the Department of Energy in 1974) began developing a series of nuclear engine projects with names like Dumbo, Kiwi, and Pluto, terminating in NERVA (Nuclear Engine for Rocket Vehicle Application). In 1960, Aerojet General Corp., Lockheed Missile and Space Co., and Westinghouse Electric Corp.'s Astronuclear Laboratory were contracted to build various elements of NERVA, including a reactor and an engine. After a dozen years, however, NERVA fizzled, and was canceled in 1972.

The most promising—or at least, the most jaw-dropping—of all the various nuclear rockets was Project Orion. Developed by General Atomics Co., a subsidiary of submarine builder General Dynamics Corp., under a contract by the US Air Force from 1958 to 1965, the Orion program was the one that got away. The one that might have been.

More than just a thought experiment with a few long and wishy-washy reports filed away, Orion was a real-world space opera on an E. E.

A part of the NERVA nuclear rocket engine being moved to a test stand in Jackass Flats, Nevada. NASA

"Doc" Smith / John W. Campbell / Poul Anderson scale. Heading the project was Theodore Taylor, who had spent eight years at the Los Alamos National Laboratory in New Mexico. An innovative manager, Theodore Taylor purposely shunned bureaucracy by adopting the management style of Germany's prewar amateur rocket society, *Verein fur Raumschiffahrt* (VfR), "Society for Spaceship Travel," whose ranks included Wernher von Braun. A model of agreeable anarchy, the VfR had no snooty division of labor, with its five hundred members as likely to work on engineering complications as they were model building. (The organization thrived until 1934, when the Nazi regime shut it down.) Taylor assembled an A-team of several scientists, many of which had worked on the Manhattan Project. Perhaps the most influential of his all-star lineup was a young Freeman Dyson, now in his nineties, and a professor emeritus in the Institute for Advanced Study in Princeton, and known for his work on crazy quilt of subjects from quantum electrodynamics to the problem of evil.

With Taylor's management approach that was more of a freewheeling playground than a corporate hierarchy, Project Orion went through several versions, each begging for new synonyms for "colossal," "supersized," and "OMFG." An early base version was a lithe, sixteen-story-high, 4,000-ton ship, with a pusher plate roughly 130 feet in diameter. The atomic bombs powering it would be 6 inches wide and weigh about 300 pounds each. Ejected at a rate of one per second, they would have yielded 0.1 kiloton. (A kiloton equals 1,000 tons of the high explosive TNT. The Hiroshima bomb equaled 15 kilotons in its explosive power.) To push this Orion into a 300-mile-high orbit, it would need eight hundred bombs released over a six-minute period. To fuel an interplanetary voyage, several thousand bombs would be required.

That was a starter-home version of Orion. Various other mission profiles were considered, including an ambitious generational ship to reach the stars. This particular one called for a 44-million-ton spacecraft housing twenty thousand people on its voyage. (The Saturn V that carried the Apollo 11 mission to the moon weighed, when fueled, about 2,800 tons.) A payload of 30 million bombs would be sequentially released in 1,000-second intervals over a 500-year acceleration and deceleration period. These were ships on a cosmic scale, so powerful they could ferry many times the amount of cargo a Saturn V could. They were so gigantic, in fact, that Ted Taylor toyed with the idea of including a 4,000-pound barber's chair on the ship—because, why not?

A subsequently slimmed down but speedier iteration of 550,000 tons would be furnished with 300,000 1-megaton bombs. Whatever their sizes, in all cases the ships' bombs' explosions and the released energies' collision with the thick pusher plate would generate an off-the-charts thrust with the force of a Lovecraftian Elder God spiking a volleyball. Dyson, who brainstormed much of Project Orion, wrote in his paper, "Interstellar Transport," that Orion would have landed men on Venus and Mars by 1968—one full year before we touched down on the moon in this reality. He was serious, too.

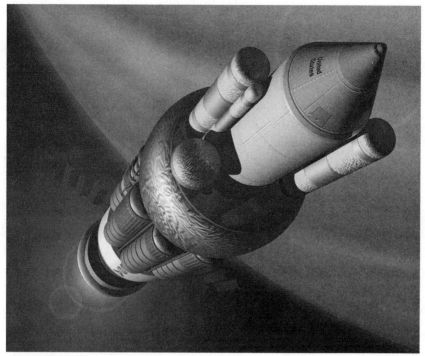

Project Orion concept drawing. NASA

In 1959 the team constructed several models, including a working 7-foot-tall scale version of Orion called Hot Rod. Using a chemical rather than a nuclear propellant, it launched the test rocket that year at Point Loma, California. During its short flight of about 300 feet, Hot Rod survived six sequential detonations. Each boom struck its scale-model pusher plate, sending the vehicle ever faster, ever higher, proving Orion was within the realm of possibility

In that moment, it must have seemed to Orion members that humanity was about to make a great leap forward, comparable to going to sleep in the Stone Age one night and waking up the next morning in the Industrial. The Saturn V—which would convey humanity to the moon—was to Dyson already in the same league as the Wright Brothers' Flyer. He sniffed at the quaint rocket, declaring it had the basic same relation to his Orion as "the majestic airships of the 1930s bore to the

Boeing 707." So marked was the difference between the Orion and its predecessors, Dyson predicted that interstellar travel, with his ships hurtling 6,200 miles per second, was even possible, allowing Orion to reach Alpha Centauri—the system whose two suns, Alpha Centauri A and Alpha Centauri B, give us an 85 percent likelihood it contains at least one habitable planet—in a speedy 130 years at 3 percent light speed.

After spending $10.4 million in developing Project Orion, General Atomics pulled the plug on the superfast ship. In a few short years, nuclear energy had gone from luminary to lowlife. The Campaign for Nuclear Disarmament, organized in 1957, helped bring about the 1963 "Treaty Banning Nuclear Weapon Tests in the Atmosphere," in Outer Space and Under Water, banning all nuclear-weapon detonations except for those held underground.

Dyson later said the technology was the first in history suppressed entirely for political reasons. Perhaps there was a pinch of truth in what he said, though that implies the fears were guided by popular hysteria instead of dispassionate wisdom. Yet in the years after the Nevada atmospheric nuclear bomb tests from 1951 to 1962, a National Cancer Institute study concluded that radioactive fallout exposed millions of American children to large amounts of radioactive iodine, which can affect the thyroid gland, with amounts ten times larger than those caused by the 1986 explosion at the Chernobyl nuclear plant in Ukraine. The study estimated the exposure was enough to result in as many as 25,000 to 50,000 cases of thyroid cancer around the country. So much radiation was splashed into the atmosphere in those days, art-fraud detectors use what's known as "bomb-curve analysis" to identify forgeries. For instance, a few years ago, the Florence, Italy–based Institute for Nuclear Physicists used a bomb-curve analysis on a disputed Fernand Léger painting from the Peggy Guggenheim Collection. Supposedly created between 1913 and 1914, the amount of carbon 14 in the painting showed it had to be an entirely postwar forgery.

In one way, Dyson's complaint is more than just sour grapes. These who fear a nuclear rocket exploding probably never gave too long a thought to what would happen if a Saturn V blew up on the launchpad. NASA did, though. In its "Saturn 5 Booster Explosion Hazards and Apollo Survivability Analyses," carried out in 1963, analysts concluded

Artist's rendition of the Deep Space 1 probe. NASA

that the rocket would explode with the force of over half a kiloton (achieving 1/30 the power of the Hiroshima bomb). The resulting fireball would be +2,498 degrees Fahrenheit and last for 33.9 seconds. That's the reason that no one is allowed within 3 miles of the launchpad.

Orion may be long gone, but the dream of hitching a faster ride to the stars remains. Electrothermal engines, for one, use a super-heated plasma, created by electrical energy, which then fire the plasma through a spout to produce thrust. About two hundred spacecraft have used this technology, including Russian satellites, since the 1970s. Lockheed Martin Space Systems' A2100 communications satellites model, built starting in the 1990s, also use electrothermal propulsion. These engines are hyper-efficient, but generate low thrust, which is why they're mainly employed in keeping satellites in their appointed orbits.

More promising is the ion drive engine, which was considered advanced even in *Star Trek: The Original Series*. In the episode "Spock's

Brain," a civilization using ion propulsion in the year 2268 shoplifts the aforementioned mind to run its government. Tantalized by arguably the most godawful episode in the *Star Trek* oeuvre, NASA researchers have tried to create an ion engine of their own. To make it work, they give a positive or negative charge (thus ionizing) to the molecules of an unreactive fuel, like xenon. Those molecules are then accelerated by an electric field that shoots out the back of a given craft, pushing it along. The thrust of an ion engine is low—so low that it's regularly been compared to the pressure a single sheet of paper exerts against your palm. What it does slowly, it also does constantly; over time, spacecraft powered by ion thrusters can eventually crank up to speeds as high as 200,000 mph. Ion engines can also supply as much as ten times the thrust per kilogram of fuel as a chemical rocket. Deep Space 1, launched in 1998, was the first NASA spacecraft to use ion propulsion, hitting speeds of 60,000 miles per hour successfully cruising asteroid 9969 Braille and the comet Borrelly.

NASA is also tinkering with Project VASIMR (Variable Specific Impulse Magnetoplasma Rocket), a plasma-based propulsion system. VASIMR uses an electric power source that in turn uses hydrogen as a rocket propellant. Because hydrogen is the most abundant element in the universe, that fact suggests a VASIMR-powered spacecraft could replenish its fuel supply . . . anywhere. Astra Rocket Co., which is currently developing the technology, says a rocket with VASIMR tech could arrive at Mars in a little more than ninety days.

By far the coolest of these is the perpetual-motion machine (or at least, a perpetual-motion machine's distant cousin), scientists at NASA's Eagleworks Laboratories are working on, called an EmDrive. It seems to treat the laws of physics as nothing more than a YIELD sign at an intersection. By bouncing microwaves around inside an enclosed cavity in the shape of a cone, the EmDrive generates thrust. There is no need for a propellant of any kind, as solar power can be used along the way in a space voyage to produce the propelling microwaves. No one is yet exactly sure why, but research from various groups testing EmDrives seem convinced that somehow, some way, it probably actually, really works. Finally, we're getting an engine as big as space itself.

CHAPTER FIFTEEN

Burning Bright: Fire Prevention

According to the National Fire Protection Association (NFPA), in 2016 there were 1,342,000 fires reported in the United States. That means roughly every twenty-four seconds, a fire department has responded to a fire. Nearly 3,390 civilians died in those conflagrations in that year, and almost 15,000 more were injured. The price of these infernos is high, including the deaths of 69 firefighters, and injuries to 62,085 more. In 2017 another 60 firefighters died on the job.

The deaths were tragic and the injuries were terrible, yet these figures also represented a steep decline over the casualties of preceding decades. In fact, from 1977 to 1991, the fewest deaths of firemen recorded in a year never went below 108, and once reached as many as 174. The 60 on-duty firefighter deaths in 2017 was the lowest total ever reported by the NFPA. There have now been fewer than 70 deaths a year in six of the past seven years.

The lifesaving changes can in part be attributed to a 1,000-second-long descent into an atmospheric Abaddon, one that was more than twice the temperature of the lava that spilled out of Hawaii Volcanoes National Park in 2018, and hot enough to reduce a hardened block of titanium to a squishy puddle.

When Apollo 11 returned to the Earth on July 24, 1969, it would have been the textbook definition of an anticlimactic denouement—especially after the first-in-history moon landing and walk, and the historic catchphrase—except for one short stretch of time when the astronauts returned to their planet of origin. Just before reentry into the Earth's atmosphere, the astronauts chucked the service module from the command module (CM). Then they deployed its thrusters to alter the vehicle's attitude—that is, the CM's orientation respective to the Earth's surface—so that its base faced toward their home planet.

The reason behind the adjustment was literally one of life and death. During a typical low Earth orbit reentry, a ship can reach a velocity of 17,500 mph, or about 3.6 times as fast as an F-35 Lightning II fighter jet at its highest *Top Gun* speed. While the ship slides down this atmospheric chute, the chemical bonds of the air's nitrogen/oxygen molecules surrounding it snap apart with the ease of breadsticks. That fissure produces an electrically charged sheath of plasma—the most common state of matter in the universe that makes up the sun and stars—around the CM, bouncing back radio signals like a basketball dropped on a hardwood floor. All communication for several tense minutes ceased.

Meanwhile, the temperature on the CM's surface scaled up to +5,000 degrees Fahrenheit. (Keep in mind, lead melts at +622 degrees Fahrenheit, and a whole turkey can be considered done when its internal temperature is about +165 degrees Fahrenheit.)

Fire knows nothing of mercy. A simple chemical reaction, fire is a process that releases light and heat. Scorched bones and charred leaf and twig fragments discovered in South Africa's Wonderwerk cave suggest that *Homo erectus* could control fire one million years ago. These early hominids' ability to understand and control fire prompted Harvard University biological anthropologist Richard Wrangham's "cooking hypothesis." His eureka notion, back in the 1990s, suggested that the doubling in human brain size, from *Homo habilis* 2.3 million years ago to *Homo erectus* five hundred thousand years later (that's an eye blink to evolution), was due to cooking skills. Essentially, his idea held that cooking, by virtue of it necessarily harnessing fire, would have permitted easier chewing and digestion, thereby making additional calories available to power energy-starved brains. If that's true, fire made us what we are today.

And yet even after twenty-three thousand centuries of conquering and yoking flame, nothing injects a lightning bolt of fear into a human being like a fire in an enclosed space, far from help, with only a fuel depot of pure oxygen. Fire transformed the space program, no less than the *Titanic* altered the maritime industry.

Vigilant to the point of paranoia against the dreaded possibility of its human cargo experiencing the same fiery demise as the astronauts in Apollo 1, NASA had fortified the CM with a heat shield designed and constructed by Avco Corp. Comprising a brazed-steel honeycomb structure infused with phenolic epoxy resin, the shield, called Avcoat, was *ablative*—meaning, to dissipate the tremendous broiling energy that accumulated during reentry, it would char, burn, and flake the cast-away chips, literally ferrying off much of the heat that built up, in order to protect its fragile inhabitants. (Coincidentally, the concept of an ablative heat shield was also put forward by rocket pioneer Robert Goddard in the

1920s. It was one of many visionary wonders he contemplated, then kept clandestine.) So vital was the heat shielding that the earliest test flights of Apollo CMs were mainly focused on testing what kind of thermal beating the heat shields could withstand.

On January 27, 1967, about two and a half years before Apollo 11 triumphantly reentered the Earth's atmosphere that July day, a little past noon, Gus Grissom, Ed White, and Roger Chaffee entered the Apollo 1 (initially designated AS-204) command module for a launch rehearsal test, for what would have been the first Apollo manned mission, scheduled for a February 21, 1967, launch. It was what was known as a "plugs-out" test, which would determine if the CM could operate on internal power while removed from all other cables and connections. The test, which had been routine since the Mercury and Gemini programs, was considered safe, since neither the launch vehicle nor the spacecraft were loaded with fuel.

Grissom, among NASA's first group of astronauts, the Mercury Seven, had become the second American in space in 1961. Fellow spaceflight veteran Ed White, an air force lieutenant colonel, was the first American to conduct a spacewalk, on Gemini 4 in 1965. Roger Chaffee was a navy lieutenant commander who joined the space program in 1963.

From the instant they settled in the CM that day at 1:00 p.m., a series of problems, none necessarily an emergency in and of itself, began. The first ensued when Grissom noticed a strange odor he described as "sour" after he hooked up to the CM's oxygen supply.

Next, a high oxygen-flow indication occasionally triggered the master alarm. Environmental-control-system personnel concluded that somehow the crew's physical movements were causing it. Last, communication glitches began cropping up, at first between Grissom and the control room; then, the malfunctions started occurring between the operations and checkout building and the blockhouse, the reinforced concrete structure where NASA personnel were housed, allowing them to monitor the launches on their equipment. Periscopes in the windowless blockhouse allowed those inside to peek out at the nearby rockets.

The communications snafus halted the countdown exercise at 5:40 p.m. Slightly less than an hour later, at 6:31 p.m., the test was about to start up again when ground instruments revealed a mysterious spike in the oxygen flow going into the space suits. Four seconds later, the words "Fire, I smell fire" floated almost casually over the intercom. Two seconds later, White declared, "Fire in the cockpit."

Deke Slayton, sitting at his chair inside the blockhouse, shouted, "What the hell's happening?"

Slayton, the Wisconsin farmer's son, was, like Grissom, a member of the original Mercury Seven group formed in 1959. He had been grounded for years because of a slightly irregular heartbeat. Now he was the director of flight crew operations, and had handpicked the team for this mission. Through a closed-circuit television monitor positioned outside the spacecraft, he could see white flames building and smoke expanding dangerously, uncontrollably.

In theory, an emergency escape routine should have occurred within a minimum of ninety seconds. In practice, the crew had never been able to perform an exit in that infinitesimal amount of time. White tried to activate a ratchet-like device that would in turn release the first in a series of latches. After just one full turn of the device, he was overcome by smoke.

Technicians lugging extinguishers sprinted to the capsule, hoping against hope they were not too late. Once they arrived, it took them about five minutes after the fire had started to open the hatch.

Slayton raced to the capsule. Fearing the worst, he looked inside to see a hellscape of black ash inside the hatch. The bodies were still, like suits with crash-test dummies in them. A medical board found that three astronauts had died of carbon monoxide asphyxia, with burns contributing to their demise. Fire had annihilated 70 percent of Grissom's space suit, 20 percent of White's, and 15 percent of Chaffee's. (No one could say how much of the burns came after the three had died. The hope that they expired swiftly from smoke became the certainty that they had passed quickly.)

The three were not the only fatalities in spaceflight programs. That same year Soviet cosmonaut Vladimir Komarov's Soyuz 1 capsule crashed onto the ground at full, unimpeded speed, killing him on contact. Komarov, who knew the craft was shoddily built (there were at least 203 known structural problems), went ahead and piloted it to save the beloved Yuri Gagarin from going in his place. He howled in impotent

The fire-wracked Apollo command module. NASA

fury all the way down, as his parachutes failed to operate. The landing was so hard, his body turned into a charred husk on impact.

Four years later cosmonauts Georgi Dobrovolski, Viktor Patsayev, and Vladislav Volkov died when the breathing ventilation valve in their Soyuz 11 craft ruptured, strangulating them. When the recovery team opened the door of capsule, they found the three with blood dripping from their ears and noses and stains of deep blue covering their faces. When the faulty valve pressure collapsed, it had exposed the doomed trio to the vacuum of space.

The first fatal accident in the history of US human spaceflight left many worried the space program would be derailed for perhaps as long as a year, as accusations of cutting corners—which would which later haunt the Challenger disaster—began circulating. NASA convened what it called the Apollo 204 Accident Review Board to determine the cause of the fire. Both houses of Congress formed committees to oversee NASA's investigation. The culprit, NASA eventually found, was electrical. The fire spread quickly due to combustible nylon material, and the pure oxygen cabin atmosphere. The astronauts' rescue was prevented by the plug door hatch, which could not be opened because of the higher internal pressure of the cabin.

Traumatic fire changes us. The 1871 Chicago fire that killed an estimated 300, left nearly 100,000 homeless, and destroyed 17,500 buildings and 73 miles of street resulted in the "Great Rebuilding" effort to create an improved and safer urban center. The Triangle Shirtwaist Factory fire in 1911 that killed almost 150 had a ripple effect extending into improved building and fire-safety codes, upgraded firefighting capabilities, and Union organizing that impelled the growth of the International Ladies' Garment Workers' Union.

Fire would change NASA, too.

In the spring of 1967, NASA announced that the mission originally scheduled for Grissom, White, and Chaffee would be known as Apollo 1, and that the first manned Saturn V launch, scheduled for November 1967, would be known as Apollo 4. (No missions or flights were ever designated Apollo 2 or 3.) The agency modified the most egregious of the Apollo spacecraft's failings, putting astronauts in fireproof suits, further fireproofing the capsules' interior, redesigning the hatch to make it easier to open from the inside, and reducing the capsule's atmosphere to a less-flammable mixture of nitrogen and oxygen in place of highly combustible pure oxygen.

As successful as it was, the Avcoat shield wasn't used for the space shuttle orbiters, but now a reformulated, ecofriendly version is scheduled to be employed for the next-generation Orion spacecraft (not to be confused with the proposed nuclear-powered spacecraft concept of the same name from the 1950s and '60s). The deep-space craft is intended for manned missions to our own moon, the Martian satellites, and ultimately, Mars itself.

The greatest impact of these heat-defying technologies wasn't necessarily up in space but down on Earth. Part of the effort to make space travel safer spun off into equipment that has saved countless lives. After the Apollo 1 deaths, NASA contracted with Celanese Corp. to develop a line of heat- and flame-resistant fabric in space suits and vehicles made of what are known as PBI textiles. The fabrics were inflammable, with no melting point, and held on to their strength and flexibility even after prolonged exposure to fire. The textiles became a part of Apollo, Skylab, and space shuttle missions. PBI fibers started filtering down to fire departments in 1978, and became commercially available five years later.

NASA financed Avco's efforts to develop improved applications of its heat shield, including fire-retardant paints and foams for aircraft. That research led to the first intumescent epoxy material, a substance that, when exposed to heat, inflates in volume, creating a kind of insulating barricade that dissipates heat through burn-off. Similar offshoots included a protective insulating layer over steel that gives the metal four hours of fire protection, in effect, holding back the fatal heat to give occupants time to flee.

Perhaps the most significant benefit of that technology went to those who combat the flames. Much of the firefighter gear that is now standard is based on NASA technology that has evolved since Apollo 1. One joint project between NASA and the National Bureau of Standards was a lightweight breathing apparatus that protects firefighters from smoke-inhalation injury.

No matter how useful firefighters' equipment may be, if its weight is prohibitive, it is little different than attaching a ball and chain around their ankles. Relieving that burden was another eventual NASA spin-off: a backpack system that carried a thirty-minute air supply. It weighed just 20 pounds, about 65 percent less than previous firefighting tanks. Equally as important as its reduced weight was the backpack's improved design, with a frame and harness that redirects the unit's weight in such a way that users carried more of its weight on their hips than their shoulders. An alarm that alerts firefighters when they're running low on air was personalized so it can only be heard by the individual, eliminating the confusion that occurred when they detected each other's alarms and mistook them for their own.

During a six-hour-long meeting on August 19, 1966, NASA officials and the Apollo 1 crew hashed out their worries over problems great and small with the Apollo 1 craft. Near the end Grissom brought the proceedings to a halt. He pulled out two copies of a photo in which Grissom, White, and Chaffee's heads are bowed, as if deeply in prayer. Grissom gave one signed copy to the Apollo program's office manager, and the other to the general manager of North American Aviation's space division, the company that had built Apollo 1. The joke-that-wasn't-a-joke implied they all knew—and accepted, with no small measure of humor and grace—how rickety and imperfect the Apollo 1 was. Grissom underscored the funereal undertone shortly before the fatal fire when he said, "If we die, we want people to accept it. We're in a risky business, and we hope that if anything happens to us, it will not delay the program. The conquest of space is worth the risk of life."

But then, as President Kennedy said in his famous "moon speech" at Houston's Rice University in 1962, "We choose to go to the moon in this decade and do the other things, not because they are easy, but because they are hard."

Moonlighting: President Kennedy addresses Congress on May 25, 1961. NASA

Lt. Col. Virgil Ivan "Gus" Grissom, 1926–1967. NASA

This "risky business" and hard things left a matchless legacy of lives saved that would otherwise have been lost. Their deliverance moderates tragedies' rough violence, and outright meaninglessness—but I think of Grissom, who seemed like an invulnerable, immortal giant to me the day I shook his hand, fifty years and more gone, anguished by flame. I think of the memorial you can find at the defunct launchpad of what was Cape Canaveral Air Force Station Launch Complex 34. It is as simple as a prayer, as brief as a dream.

In memory of those who made the ultimate sacrifice
so others could reach for the stars
Ad astra per aspera (a rough road leads to the stars)
God speed to the crew of Apollo 1

CHAPTER SIXTEEN

Blowing Up: Moon Bombs

IT IS EASY TO FORGET THAT NUKES AND THE EARTH'S SOLE SATELLITE once went together like the kiss-on-the-cheek rhyme of moon and June and spoon. Many today look back on the Cold War era and remember it as one where, in 1962, Earth was "minutes" away from "catastrophe," as a Soviet general later said.

He was referring, of course, to the 1962 Cuban Missile Crisis, when, after a U-2 reconnaissance plane spotted Soviet missile bases in Cuba, the United States and the Soviet Union played a game of nuclear chicken. Following thirteen tense days of negotiation and bluffing—including President Kennedy ordering a naval blockade around the northern Caribbean island—Soviet leader Nikita Khrushchev publicly agreed to remove Soviet nuclear missiles in Cuba, and President Kennedy secretly promised to withdraw intermediate nuclear missiles from Turkey. (Bonus: No Cuba invasion.) "We're eyeball to eyeball," Secretary of State Dean Rusk said, "and I think the other fellow just blinked."

Memories were still strong then of the photographic image snapped by Lt. Charles Levy of the 45,000-foot-tall mushroom cloud levitating over Nagasaki in 1945. "It was purple, red, white, all colors," Levy said of the 20-kiloton weapon's aftermath, "something like boiling coffee. It looked alive . . . we were all plenty scared." Imagine the scenario of that destroyed city multiplied by the 18,638 nuclear weapons the United States had stockpiled by 1960, ready for whatever rough beast slouched toward us from Moscow. Any photographs taken of that wreckage would have captured a death row infinitely long, mausoleums of sand and ash.

Mushroom cloud cake. LIBRARY OF CONGRESS

Even popular culture squirmed with anxiety then, its movies glowing with cinematic angst awakened and empowered by nuclear radiation. Eight-foot-long ants in *Them!* scuttled in the public mind the same year the 164-foot-tall prehistoric sea monster Godzilla stomped into it. Egging the dread on were *The Incredible Shrinking Man* alongside *Attack of the 50 Foot Woman*, their titular characters emasculated and defeminized by radioactivity.

Nuclear didn't always mean bloodcurdling. Sometimes it meant dead sexy. One of the first nuclear bombs dropped on the Bikini Atoll Marshall Islands was rumored to have Rita Hayworth's image painted on it. Footage from Robert Stone's 1988 documentary *Radio Bikini* suggests only her name was stenciled on the explosive, but the *Kama Sutra*–esque coupling of nukes and sex that began with that simple act of lettering had already reached a point of no return.

Literally days after the first mushroom cloud floated over Hiroshima on August 6, 1945, the Burbank Burlesque Theatre in Los Angeles

Mushroom cloud over Bikini Atoll. LIBRARY OF CONGRESS

invited citizens to come see the svelte "Atom Bomb Dancers." After the Bikini Atoll Marshall Islands test in 1946, French designer Jacques Heim, known for innovating with fur and beach outfits, dropped the first nuclear couture with his "Atome," hiring a skywriter to promote it with an airy dispatch, calling it the "world's smallest bathing suit." Not to be vanquished by a competitor whose supposedly "risqué" suit still left the innocent navel covered in undeserved mystery, Louis Réard, a onetime automobile engineer, came out with the bikini, named for the site of the tests conducted only a month earlier.

Exposing his rival's inflated claim to salaciousness, Réard in turn advertised his swimsuit as "smaller than the world's smallest bathing suit." Nineteen-year-old Micheline Bernardini, a nude dancer from the Casino de Paris, "a hotbed of extravagant costumes and topless dancers," was the only model willing to wear what amounted to 30 inches of material in the bikini's public debut at the chic Piscine Molitor in Paris. A hotel/swimming pool complex designed to resemble an art-deco ocean liner, Piscine Molitor was the place in Paris to see and be seen.

Posing for photographs in her semi-exposed state, Bernardini held up a matchbox in her left hand, symbolizing how little storage space was needed to roll up and store the minuscule swimsuit. A born salesman, Réard issued ads declaring that no two-piece swimsuit could be considered a true bikini "unless it could be pulled through a wedding ring."

Reard wasn't the only one whose fortunes were increased by the released atom. In a 1946 issue, Superman, in order to save his paramour Lois Lane, gulps down a poison that renders him momentarily senseless. In his addlepated stupor, and with the reasoning of a meth head, the Man of Steel speeds directly into the atomic bomb test at the Bikini Atoll, which providentially renders him lucid again. Atomic everything was the new normal, from the *A is for Atom* alphabet book to the Gilbert No. U-238 Atomic Energy Lab, the nuclear physics educational set that came with a Geiger counter, a "Prospecting for Uranium" booklet, and four glass jars containing radioactive uranium-bearing ore.

Nevada casinos capitalized on the nuclear tests that began in the state in 1951 at the Nevada Proving Grounds, the 1,375-square-mile site near the Nellis Air Force Gunnery and Bombing Range in Nye County, roughly 65 miles from Sin City. The first test, nicknamed "Able," at the Proving Grounds was followed by about one hundred more atmospheric nuclear trials at the site. Packing SRO crowds on their roofs and in their penthouse suites as primo seats from which to see the bombs' glowing funnels and radioactive puffballs, the casinos offered special "atomic cocktails" and "Dawn Bomb Parties." To help every visitor get in on the

Tourists could view atomic blasts from 65 miles away in downtown Las Vegas.
LAS VEGAS NEWS BUREAU

boom, the Nevada Chamber of Commerce handed out calendars listing the upcoming schedule of blasts.

Soon even beauty pageants were produced to crown "Miss Atomic," "Miss Atomic Bomb," "Miss Atomic Blast," and "Miss A-Bomb." Arguably the most famous was Miss Atomic Bomb 1953, Copa showgirl Lee Merlin, who wore fluffy cotton mushroom clouds attached to the front of her swimsuit. Her arms aloft, with a 1,000-watt smile, she could be ringing in New Year's, or writhing in a frenzy of erotic excitement.

With the 1955 US Atomic Energy Commission brochure that assured readers that "The path of fall-out . . . does not constitute a serious hazard to any living thing outside the test site," the government went on to conduct 928 nuclear tests (most of them underground). Over the next four decades, the Nevada Proving Grounds earned the nickname "the

Miss Atomic Bomb 1953, Copa showgirl Lee Merlin, in her fluffy mushroom-cloud costume. LAS VEGAS NEWS BUREAU

most bombed place on Earth." The angle, a Nevada official was readily admitted, was to get people to think "the explosions wouldn't be anything more than a gag." Nevada's Clark County (which includes Las Vegas) even changed its official seal to include a giant mushroom cloud (it now bears the picture of a Joshua tree). Still, the omnipresent cloud—the Nike swoosh of its time—even appeared on the Clark County telephone directory and *The Wildcat Echo*, a 1953 Las Vegas high school yearbook.

With the atom and erotica joined like coffee and Danish, it isn't a stretch then to see that the space race was marked by an unhinged optimism about nukes, just as we once thought the Internet would be an unambiguous force for democracy, toppling tyrants, and letting every suppressed voice be heard.

In 1958 senior US Air Force officers approached Dr. Leonard Reiffel, a physicist at the Armour Research Foundation at the Illinois Institute of Technology, to "fast-track" a project to detonate a nuclear bomb on the moon. The resulting top-secret Project A119 was explored in a blandly titled 1959 report called "A Study of Lunar Research Flights." The report, disclosed in 2000, may be the sole remaining declassified document relating to Project A119. It's likely that Reiffel produced eight other reports on the plan's progress, all of which have likely been destroyed and lost to history.

The remaining 1959 study is an Encyclopedia Britannica of everything we knew about the moon back in the 1950s, from its geology to its atmosphere. All the information in its 190 pages pulses with Ahab-like diligence to answer one simple but imperative question: What kind of nuke would be visible from that distance?

"It was clear the main aim of the proposed detonation was a PR exercise and a show of one-upmanship," Reiffel later recounted. The Air Force wanted a mushroom cloud so large it would be visible on Earth, because the United States was lagging behind in the space race. "Lagging" seems almost like a participation-trophy kind of word for the United States then. The Soviets dominated space as much as it once did in chess. The USSR's winning streak of firsts was a public rebuke to the American way:

- First to launch an artificial satellite (Sputnik, 1957)
- First to put an animal in orbit the (Laika, 1957)
- First flyby of the moon (Luna I, 1959)
- First spacecraft to land on the moon (Luna II, 1959)
- First to land on another planet (Venus, 1961)
- First to put a man in space (Yuri Gagarin, 1961)
- First to put a woman in space (Valentina Tereshkova, 1963)
- First to perform a spacewalk (Alexey Leonov, 1965)

For the United States, each milestone felt like a hearse pulling closer and closer to the front door. The mind-set was eloquently voiced by Senator Lyndon Johnson in 1958. Johnson, the Majority Leader of the Senate, also chaired the Senate Armed Services Committee's Preparedness Subcommittee, where he would target the freshly coined concept of "The Space Race." In a summary of the subcommittee's report, Johnson may as well have called his position "Manifest Destiny 2.0": "Control of space means control of the world," Johnson said. "From space, the masters of infinity would have the power to control the earth's weather, to cause drought and flood, to change the tides and raise the levels of the sea, to divert the Gulf Stream and change temperate climates to frigid."

Because paranoia is security's go-to viewpoint, the United States assumed the Soviets were brewing a similar scheme, especially after the *Pittsburgh Press* reported on November 1, 1957, that "The latest rumor going the rounds is that the Russians plan to explode a rocket-borne H-bomb on the moon on or about November 7," to celebrate the fortieth anniversary of their revolution.

The Soviets didn't commemorate the revolution in 1957 with a nuclear shot seen round the world. However, in 1958, made confident by the launch of Sputnik the year before, two Russian scientists, Sergei Korolev and Mstislav Keldysh, developed the four-step "E Project." Based on the earlier work by the accomplished and highly respected Soviet nuclear physicist Jakov Seldovich, E-1 involved simply getting a spacecraft to the moon. The next two stages, E-2 and E-3, were slightly more ambitious, orbiting the moon and photographing its surface. E-4, however, took a leap that was as risky as driving drunk into a minefield: detonating a small nuclear charge on the lunar surface.

Seldovich's original idea was levelheaded enough: Drop a nuke on the moon's surface to provide the world with irrefutable proof that the Soviet Union had reached it. It made sense, because a spacecraft itself would be too tiny to detect by any Earth-based astronomer, and any declaration by the Soviets of such a gigantic accomplishment might be lumped with the long history of Russian shams on a grand scale. Brazen trickery, like Grigory Potemkin's alleged pasteboard facades of pretty towns with well-dressed serfs, installed to convince Catherine the Great

AFSWC-TR-59-39

SWC
TR
59-39
Vol I

HEADQUARTERS

AIR FORCE SPECIAL WEAPONS CENTER

AIR RESEARCH AND DEVELOPMENT COMMAND

KIRTLAND AIR FORCE BASE, NEW MEXICO

A STUDY OF LUNAR RESEARCH FLIGHTS

Vol I

by

L. Reiffel

ARMOUR RESEARCH FOUNDATION

of

Illinois Institute of Technology

19 June 1959

the peasantry was prosperous, and famine wasn't about to scythe through her subjects.

A nonworking mock-up of the spacecraft was made. The mock "nuclear device" was constructed with initiator rods set in multiple directions, much like an antishipping mine, to ensure an explosion at the moment any part of the rods touched the moon's surface. Once it blew and the world took notice, it would be tantamount to the USSR chalking the moon with graffiti, saying "The Kremlin, not Kilroy, was here."

Eventually the plan lost its backing, especially from Seldovich himself. The bomb might boomerang during launch, and instead of lighting the moon, end up burning a hole in the USSR, or, worse, another country. Dampening the initial fervor even more was the realization that the flash would be short-lived because of the lack of an atmosphere on the moon, and therefore might not even register on film.

Meanwhile, back in the United States, Reiffel, with the assistance of ten researchers, moved ahead on A119. Among his talented team was Carl Sagan, the beloved popularizer of science, who Reiffel hired to mathematically model how visible the exploding dust cloud would be from the Earth.

The conclusion, meticulously measured and methodically evaluated, was that the explosion would be best staged on the moon's dark side, with the desired mushroom cloud floodlit by the sun. At the outset, the program's participants considered a hydrogen bomb, since that's what the Soviets were apparently planning to use. However, the air force rejected that out of hand: A hydrogen bomb would be far too heavy to launch into space. Instead, a smaller nuke, one which would burst with 1.7 kilotons of force, would be light enough and do just fine. Compared to the 15-kiloton bomb that fell on Hiroshima, this would be a firecracker.

Like the Soviets, the United States rejected the idea over the anticipated public outcry that would result from putting nuclear weapons in space, an uproar that would no doubt escalate when the idea that an armed rocket might malfunction and crash back to Earth sunk in. It gave the project not just a deadly chill but also a fatal pneumonia.

A119 was just one of many military efforts aimed at the moon. The army's Project Horizon would have put a military base on the moon at a

HERE MEN FROM THE PLANET EARTH
FIRST SET FOOT UPON THE MOON
JULY 1969, A. D.
WE CAME IN PEACE FOR ALL MANKIND

NEIL A. ARMSTRONG
ASTRONAUT

MICHAEL COLLINS
ASTRONAUT

EDWIN E. ALDRIN, JR.
ASTRONAUT

RICHARD NIXON
PRESIDENT, UNITED STATES OF AMERICA

"We came in peace for all mankind." NASA

cost of $6 billion, spread out over several years. Kicking off in 1959, the 118-page secret Project Horizon study was submitted to the Department of the Army 1959. One passage makes its motivation clear and concise.

> *[To] be second to the Soviet Union in establishing an outpost on the Moon would be disastrous to our nation's prestige and in turn to our democratic philosophy.*

As the plan progressed, it used near-term projections with available or almost-available technology to make its case: Launch five Saturn rockets every month until "490,000 pounds of useful cargo" had been dropped on the moon. Those shipments would be followed by 252 men placed in Earth orbit by the close of 1967, with 42 of them continuing on to the moon for their one-year tours of duty. But the resource drain of the Vietnam War and the 1967 "Treaty on Principles Governing the Activities of States in the Exploration and Use of Outer Space," including the Moon and Other Celestial Bodies, which prohibited establishing military bases on the moon, slowed the project to a dead stop.

"Everyone is a moon, and has a dark side which he never shows to anybody," Mark Twain once wrote. A119 and the E Project were humanity's murky side. Despite the technology, these schemes and intrigues were as old as papyrus, and now they're as lost as Atlantis. At least the stainless-steel commemorative plaque Apollo 11 left behind on the moon could say, without a smirk of irony, "We came in peace for all mankind."

CHAPTER SEVENTEEN

Tooling Around:
Dustbusters

THE SPACE RACE WAS STILL IN ITS EARLY LAPS, AND NASA WAS SPRINT-ing like a runner who hadn't heard the starting pistol go off. Among its seemingly endless needs was for a drill that could operate in conditions no human—or implement, for that matter—had ever attempted. For one, the Apollo astronauts were going to collect samples of lunar rock and soil on the moon. Much of them they could likely just reach down and pick up from the lunar surface, like any rock hound casually hiking a nature trail (admittedly, a nature trail without the convenience of breathable air and where the temperature dips to a frosty -280 degrees Fahrenheit), assuming the gloves were supple enough to handle that particular chore. But to acquire a broad and extensive sampler of moon geology, astronauts might also have to drill down as much as 10 feet beneath the surface. This was going to be a little more difficult than drilling a pilot hole in a 2-by-4, for example, or installing drywall for a new rec room.

First off, the drill would have to be powerful enough to buzz through a lunar surface whose hardness might either be off the scale, or soft and easily penetrated—no one could really be completely sure. The best guess, based on the limited data taken from the Americans' Ranger 7 and 8 and the Soviets' Luna 12 and 13 unmanned probes, was that astronauts could expect to encounter a layer of loose debris. This rubble—i.e., dust, soil, broken rock—known as *regolith*, could be several feet deep. The moon's regolith (the word is a blend of two Greek words, *rhegos*, which means "blanket," and *lithos*, which means "rock") was formed over 4.51 billion years ago by a relentless barrage of meteorite impacts on its surface.

Nobody was entirely sure, but the floor mat of fragmented rock and other related materials might run anywhere from 10 to 66 feet deep. That's a nervously large margin of error.

If the drill Apollo crews took with them wasn't up to the task, they could just return it to a hardware store and swap it out for new one. The tool would have to be lightweight and compact, as well as powerful enough to cut through the possibly sturdy lunar surface layers. Additionally, it would have to be cordless. While the lunar module probably had enough power to run a cord from it to the drill, the missions would require a sampling of rocks from an assortment of locations, and some of those would only be reached by cruising out to them in the lunar rover. Thus, that option was ruled out. Besides, cords and outlets were problematic in a way far beyond the typical Earthly annoyances they cause. They might fray, they can get tangled up, and if the lunar module outlet were damaged in some way, the closest electrical outlet would be several days' flight away.

Isaac Newton was also going to be a problem. More precisely, the seventeenth-century polymath's Third Law, popularly understood as "For every action, there is an equal and opposite reaction." Example: A Saturn V rocket exercises a massive backward force on the gases its engines expel at launch, and those gases consequently apply an equally massive reaction force on the rocket, pushing it up, up, up in the heavens.

Newton's Third Law meant torque—a twisting force that tends to cause rotation—was another potentially disastrous complication. An astronaut using a high-torque drill in space could end up being spun around the drill bit as he tried to twist it, thus slinging him into the space above like a Frisbee. Not the optics NASA's PR department was likely hoping for.

NASA turned the job of developing the drill over to the Black & Decker Manufacturing Co. The company already had a successful track record; a few years before, in the mid-1960s, Martin Marietta Corp. had subcontracted with Black & Decker to design a wrench for the Gemini missions that could manage to spin bolts in zero gravity without inconveniently spiraling the astronaut using it, as well. Project Gemini astronauts first used the cordless Minimum Torque Reaction Space Tool in 1964 without any known problems of a Newtonian nature. On top of that successful entry on its résumé, Black & Decker had also introduced the world's first cordless rechargeable drill in 1961.

The true evolution of power tools that went to the moon began in 1916, with the portable electric drill's invention. A little more than a century ago, Black & Decker filed a patent application for a half-inch portable drill that could be operated by a single person. Its design made an AK-47 look frilly by comparison: a universal electric motor, able to run on alternating current or direct current, and a pistol-grip handle with a trigger control. If it looked like a nonlethal gun, that's because its inspiration came directly from firearms. Owners of a machine shop in a Baltimore warehouse, S. Duncan Black and Alonzo Decker counted among their clients the gun manufacturer Colt. One day, stumped over the most efficient way to hold the tool and control the drilling function at the same time, their attention turned to a Colt handgun lying nearby.

Astronauts experience weightlessness—and nausea—on the "Vomit Comet" NASA

The firearm's simple pistol grip and trigger inspired the template for their drill, and dozens like it ever since.

The company developed a battery-powered system that could operate smoothly in the stressed conditions of extreme temperatures and zero atmosphere. But before the zero-impact wrench and rotary hammer drill went into space, they had to be tested in conditions simulating low gravity, by trying them out on the so-called "Vomit Comets," the planes that climb to a high altitude then quickly nosedive at a nausea-inducing angle. The aircraft's deep, downward plunge typically creates about twenty-five seconds of complete weightlessness to work in. (Another angled dive can produce a lunar-like one-sixth Earth gravity, lasting about forty seconds.) Because the window of opportunity is so short, the steep upward / steep downward maneuver has to be repeated over and over to thoroughly put any equipment through its paces. (First started by the US Air Force in 1957, what became known as the Reduced Gravity Program was taken over by NASA in 1973, which managed it until 2008. Since that year, the undulating and sometimes stomach-churning plane ride has been operated by Zero Gravity Corp., using a modified Boeing 727.) Anyone can tag along, provided they have about $5,000 of disposable income on hand for the trip.

Ultimately the lunar surface drill was first used on Apollo 15 in 1971, when David Scott and James Irwin drilled below the lunar surface and collected a subsurface core on the way to amassing about 170 pounds of lunar surface material. In all, the Apollo missions hauled back a cargo of almost 880 pounds of 400 lunar samples.

The lunar drill may have worked in a vacuum, but it didn't always exist in one. Black & Decker created several spin-offs from the technology, including the lightweight battery-run Mod 4 Power Handle Cordless System. A kind of multipurpose tool, the Mod 4 was a grass shear that could, with the addition of various extensions, transform itself into a drill, a shrub trimmer, a sealed-beam lantern, and a small vacuum cleaner called the Spot Vac.

Introduced in 1974, the Mod 4 found acclaim in the pages of *Fortune* magazine, which lauded it as one of the country's twenty-five best-designed products in May 1977. Despite a bounty of similar accolades and tributes, the Mod 4 flopped.

James Irwin scoops up lunar soil for the Apollo 15 mission NASA

The company's market research was able to extract one useful nugget of gold from all its commercial compost: 92 percent of females who used the Mod 4 particularly liked the Spot Vac's vacuum function. Over the next few years, Black & Decker deconstructed the Spot Vac and redesigned it, adding a variety of colors, revising the charger shape, and downplaying the brand name. It changed the name to the strikingly memorable "Dustbuster," after the winning entry in a company-wide contest. Debuting at the 1978 Hardware Show in Chicago, the convenient household device single-handedly created the handheld vacuum market, which grew 300 percent between 1981 and 1984.

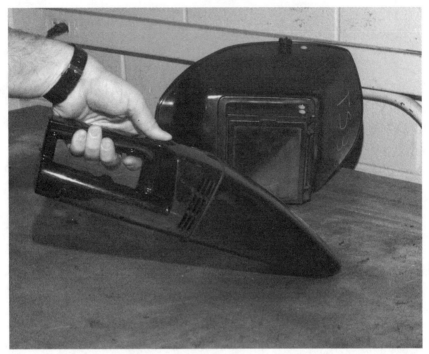

The household tool from outer space: Dustbuster. NASA

By 1995, 100 million Dustbusters had been sold. Recognizing the imprint it made on the cultural scene, the Smithsonian placed an original model of the 1979 Dustbuster in the National Museum of American History. There it took its rightful place alongside an 1898 Kodak camera and one of Isaac Singer's early sewing machines.

Chapter Eighteen

Crash-Proofing:
Memory Foam

Two words you never want to hear, speak, or read when you're about to land on a desolate object 250,000 miles from the nearest first responder, where the thermometer swings manically between extremes of +260 degrees Fahrenheit above zero Fahrenheit and –280 degrees Fahrenheit below zero: *program alarm*.

On July 20, 1969, just four minutes after Apollo 11's lunar module, the Eagle, had disengaged from the command service module, aka the Columbia, and entered into its landing sequence for the moon, Mission Control heard Neil Armstrong speak from a quarter million miles away. The bafflement in his voice somehow lost none of its tension, despite the distance. "It's a 1202 . . . What is that? Give us a reading on the 1202 program alarm."

The puzzling digits had begun appearing on the Eagle's computer display. Controllers in Houston scrambled through their notes to figure out what exactly a 1202 was. But time was running short. Eight minutes remained until the Eagle would meet the surface of the moon one way or the other.

Even before the alarm, the risk factor on the final stages of the moon landing had begun to climb to point-of-no-return levels. The radio communications link had become scratchy, effectively choking off Mission Control's ability to assess whether the Eagle could safely land. Gene Kranz, the flight director for this first lunar landing, understood the gravity of the moment. Armstrong would likely crash or call the landing off.

Apollo 11 lunar module ascent stage, photographed from command module. NASA

The former would have been a catastrophe; the latter, a debacle. Fifty years after the moon landing we remember it as a linear progression down a smooth road with few detours, speed bumps, or potholes. We forget, the way we erase a departed friend's faults, all the test rockets that exploded on launchpads, the string of second-place finishes in space-exploring milestones, the hatch malfunctioning on Gus Grissom's Liberty Bell 7 that almost resulted in his drowning, and the Apollo 1 fire that consumed three lives, including Grissom's. We remember instead a golden era in the American century when the can-do spirit dreamed big and accomplished more. The space shuttles Challenger and Columbia pricked the bubble of that illusion to some extent, but even as we were made wiser by those tragedies, we mostly recall the public record of the space race's triumphs, like the 96-point type for the *New York Times* headline, MEN WALK ON MOON, and President Nixon's phone call of congratulations to "Tranquility Base."

On June 13, 1969, not long before Apollo 11 lifted off, Frank Borman phoned President Nixon's speechwriter, William Safire. Borman, who had commanded the Apollo 8 mission, the first manned spacecraft to leave low Earth orbit and the first to circle the moon, urged Safire to have a response ready if the landing was less than 100 percent successful. "You'll want to consider an alternative posture for the president in the event of mishaps," said Borman, whom NASA had assigned to be its liaison between the agency and the president.

Safire got to work, preparing a morbid tribute for what was supposed to be one of history's greatest and happiest moments. Since the alternative statement was never used, it went into the junk drawer of our national memory, the National Archives, where *Los Angeles Times* reporter Jim Mann discovered it while researching a book about the US policy toward China. He wrote about his unexpected find in a 1999 *Los Angeles Times* article, "The Story of a Tragedy that Was Not to Be."

Once it was determined that the Apollo crew were lost in space, perhaps, or that Armstrong and Aldrin had been stranded on the surface of the moon, President Nixon would telephone each of the widows-to-be, offer his condolences, and prepare them for what he was about to tell the world.

IN EVENT OF MOON DISASTER:

Fate has ordained that the men who went to the moon to explore in peace will stay on the moon to rest in peace.

These brave men, Neil Armstrong and Edwin Aldrin, know that there is no hope for their recovery. But they also know that there is hope for mankind in their sacrifice.

These two men are laying down their lives in mankind's most noble goal: the search for truth and understanding.

To   :   H. R. Haldeman

From: Bill Safire   July 18, 1969.

- -

IN EVENT OF MOON DISASTER:

Fate has ordained that the men who went to the moon to
explore in peace will stay on the moon to rest in peace.

These brave men, Neil Armstrong and Edwin Aldrin, know
that there is no hope for their recovery. But they also know that there
is hope for mankind in their sacrifice.

These two men are laying down their lives in mankind's
most noble goal: the search for truth and understanding.

They will be mourned by their families and friends; they
will be mourned by their nation; they will be mourned by the people of
the world; they will be mourned by a Mother Earth that dared send two
of her sons into the unknown.

In their exploration, they stirred the people of the world to
feel as one; in their sacrifice, they bind more tightly the brotherhood
of man.

In ancient days, men looked at stars and saw their heroes in
the constellations. In modern times, we do much the same, but our heroes
are epic men of flesh and blood.

The greatest speech never made. NATIONAL ARCHIVES

They will be mourned by their families and friends; they will be mourned by their nation; they will be mourned by the people of the world; they will be mourned by a Mother Earth that dared send two of her sons into the unknown.

In their exploration, they stirred the people of the world to feel as one; in their sacrifice, they bind more tightly the brotherhood of man.

-2-

Others will follow, and surely find their way home. Man's search will not be denied. But these men were the first, and they will remain the foremost in our hearts.

For every human being who looks up at the moon in the nights to come will know that there is some corner of another world that is forever mankind.

PRIOR TO THE PRESIDENT'S STATEMENT:

The President should telephone each of the widows-to-be.

AFTER THE PRESIDENT'S STATEMENT, AT THE POINT WHEN NASA ENDS COMMUNICATIONS WITH THE MEN:

A clergyman should adopt the same procedure as a burial at sea, commending their souls to "the deepest of the deep," concluding with the Lord's Prayer.

The greatest speech never made (*continued*)

In ancient days, men looked at stars and saw their heroes in the constellations.

In modern times, we do much the same, but our heroes are epic men of flesh and blood.

Others will follow, and surely find their way home. Man's search will not be denied. But these men were the first, and they will remain the foremost in our hearts.

For every human being who looks up at the moon in the nights to come will know that there is some corner of another world that is forever mankind.

After the president's last somber words, NASA would shut down communication with the men, while a clergyman would employ the same protocol as a burial at sea, commending the astronauts' souls to "the deepest of the deep," and, finally, closing with the Lord's Prayer.

Its emotion lit a match in the dark. Its beats as slow as a dirge, Safire's speech has been described as "the greatest speech never given." He had drawn inspiration from Rupert Brooke's poem on World War I, "The Soldiers," a crisp salute to the fallen Englishmen whose bodies were left far from their homes on foreign soil. Even one hundred years later, its heart beats loud.

> If I should die, think only this of me:
> That there's some corner of a foreign field
> That is for ever England.

The pressure kept building aboard the Eagle. Another alarm blared. The 8,650-pound lunar module kept descending. And on the ground below, which was supposed to be runway-smooth, loomed a cratered field and boulders as big as pickup trucks. With sixty seconds of fuel remaining and the gas tank emptying fast, Armstrong fixed on a smooth spot ahead and steered the ship toward it. With thirty seconds of the dwindling fuel supply left, the ship made it down to the surface intact. "Houston, Tranquility Base here . . . the Eagle has landed," Armstrong said, in his trademark laconic tone.

There was no crash. But had there been, would the shock have killed or injured the astronauts beyond their ability to pilot the ship back to Michael Collins and the waiting Columbia command module? The possibility brings to mind Chuck Yeager's famous swipe, his "Spam in a can"

Neil Armstrong: Mission accomplished. NASA

comment about what the astronauts essentially were—meat stuffed in metal containers shot into space at record speed.

No one will know because the landing was about as soft as it could be under the circumstances. But what we do know is that years before, NASA's concern over a hard landing prompted the agency to deliver the padding that would someday become memory foam.

In the early years of the space program Charles Yost, an aeronautical engineer with the Systems Dynamics Group at North American

Nixon chats with Armstrong and Aldrin on the moon via telephone-radio transmission. NASA

Aviation, worked on making sure Apollo command modules and their contents of astronauts could land with minimal jolts and shocks to the body. A few years later, NASA contracted Yost, through Stencel Aero Engineering Corp., to develop airplane seating that could withstand intense vibrations and even severe crashes by increasing both passengers' comfort and their chances of survival in, as Frank Borman might have put it, mishaps.

Working with NASA's Ames Research Center, and researcher Chiharu Kubokawa, Yost devised a type of plastic foam that could absorb vast impacts, and then miraculously revert to its original shape as if nothing had ever happened. Technically, if inelegantly, it was called an open-cell, polymeric foam material with viscoelastic properties. The heat-sensitive foam softened from warmth wherever a human body touched it, but remained consistently firm in the cooler parts where it hadn't. In the abstract, that meant it would comfortably mold to any individual's body, allotting their weight evenly over its surface. In real-world applications, that meant ensuring a cozy ride from Milwaukee to Moscow, or someday, Mars.

Its comforting powers were surpassed only by its protective abilities. After testing several seats made of the shock-absorbing foam at Oklahoma City, Oklahoma's Civil Aeromedical Institute, researchers determined that the foam could survive a crash far better than the aircraft itself. Just a 3-inch-thick pad of the foam could absorb the impact of an average-size adult's 10-foot fall.

Despite its obvious advantages, the foam took a long and sometimes winding path into mainstream acceptance. Back in 1967 Yost formed Dynamic Systems Inc. to market the novel invention he called, somewhat clumsily, "slow spring back foam." Happily, he changed the name to the much more marketing-savvy "Temper Foam," referring to its innate sensitivity to body temperature. Meanwhile, over the years, its inherent ability to spring back to its original shape, no matter how hard or how long it's been squished, prompted people to start using the more-colloquial term "memory foam" when referring to any brand using the stuff, much as we use "Kleenex" to refer to any kind of facial tissue.

Since then, memory foam has shaped its way into every aspect of American life. Yost's company applied the spongy material to ejection seats, ski boots, wheelchair cushions, X-ray table pads, off-road vehicle seats, and even sports equipment, like baseball chest protectors and shin guards. One of its highest-profile uses was for football helmets, minimizing the blunt-force trauma of the thirty or more hits typically inflicted on players during a game. The foam absorbed blows comparable to crashing your car into a wall going about 30 mph.

In the early 1980s, NASA released the formula for viscoelastic material into the public domain. That single act resulted in memory foam becoming so pervasive that it paradoxically became almost invisible. The foam was integrated into dozens of everyday consumer products, the most conspicuous being mattresses—spearheaded by Tempur-Pedic International Inc., which likely enjoys the highest name recognition of all such bedding makers. The auto industry added it to vehicles in the Indy Racing League, NASCAR and Formula 1, and the Champion Auto

Racing Team, supplying them with foam inserts that vastly improved drivers' chances of walking away from a high-speed crack-up. Motorcycle makers worked it into their cycles' seats to help compensate for their products' sometimes-skimpy shock absorbers and lumbar supports, making those long rides on the open road more comfortable.

Likewise, a business in Kentucky built memory foam into its horses' saddles to relieve riders' discomfort. One Colorado firm used the material to build inflatable bumper rafts, which resist sinking, for whitewater rides at theme parks. Memory foam insoles are commonplace in shoes, making walking less arduous. Hospitals typically use memory-foam mattress pads and wheelchair seats to help patients avoid pressure ulcers, aka, pressure sores, decubitus ulcers, and bedsores. Tempur-Pedic's Tempur-Medical brand reports that using memory foam lowers the rate of these bedsores among patients to 1 percent or less.

The military also employed memory foam to improve shock absorption in its vehicles and to lessen the impact of incoming projectiles in its bulletproof vests.

Memory foam continues to be used in new ways because it continues to improve. Mattress makers, for example, are innovating with airflow layers and materials, such as DuPont's Coolmax fabric, which wicks sweat away. Even the brief time memory foam takes to revert to its original contours, no matter what shape it was squashed into, has been shortened.

For all of its plasticity, memory foam still retains the impression of its first nascent years guarding astronauts against fatal impacts. When these space travelers come back after extended stays aboard the International Space Station, they readapt to Earth's hard tug of gravity by working out on an obstacle course. By watching the crews walk on the course's unbalanced memory-foam floor, NASA physicians can evaluate how well these returning rocketeers are regaining their equilibrium, while the 4-inch-thick base ensures that any stumbles are trivial.

CHAPTER NINETEEN

Dressing for Success:
Space Suits

SHOOTING HUMANS INTO OUTER SPACE IS ONE OF THE MOST UNFRIENDLY things you can do to them.

Think of it this way: Imagine diving deep into the ocean. The pressure increases about one atmosphere for every 10 meters of water depth. So, at a depth of 16,400 feet, in the section called the Abyssal Zone, pressures range around a crushing five hundred times greater than that experienced at sea level.

Sunlight has given up reaching this depth, so many of the creatures found here are therefore blind. And swimming in a world whose temperature drifts a notch or two above freezing is for beasts like the viperfish, the sea dweller that freakishly resembles the xenomorph from *Alien*. Imagine abducting this ugly bit of chum from that refrigerated midnight of a world and dumping it in Aswan, located in the south of Egypt, which receives just .04 inches of rain per year, and where temperatures frequently reach three times that of the Abyssal Zone. Good times.

It would be just as comfortable for a human chucked into outer space—which, for the sake of argument, we'll say starts at 100 kilometers (approximately 62 miles) above the Earth's surface. This invisible border is officially approved by the Fédération Aéronautique Internationale (FAI), the international governing body of air sports founded in 1905. (There are clashing opinions battling it out now for where space begins. Some contend it's 73.3 miles, and some argue for a whopping 13 million miles. Peculiarly, the FAI awards "astronaut" status to anyone who flies above 80 kilometers, or about 50 miles high.)

Called the Karman line, the 63-mile border is the point at which an aircraft can't fly fast enough to generate lift for its wings. More relevant to wingless human anatomy, it's also the point at which the gas combination we breathe to live—about 78 percent nitrogen and 21 percent oxygen—is about to begin to separate because of how gravity affects these gases at that height.

It's not exactly a picnic below the Karman line, either. Once humans ascend above 20,000 feet, they need a special breathing apparatus to live. Between 40,000 and 50,000 feet, they need oxygen fed to them under pressure via a breathing mask or helmet. Above 50,000 feet, humans need to be secured inside a pressurized environment. At those altitudes, your lungs not only struggle to inhale, but also fight to expel carbon dioxide.

Then, at around 5 miles, 63,000 feet, the rising body hits the Armstrong line. At this imaginary demarcation, the atmosphere retains just 1/20th of the pressure it possesses at sea level. Here, water boils at the temperature of the human body. Here, a pressure suit becomes literally vital. Without one, the body would starve of oxygen, lose consciousness in fifteen seconds, and die roughly five seconds later.

Nothing manned at these heights would be possible without space suits.

Space suits (the word first appeared in 1962), whether they're the primitive and stiff getup from the early Mercury program, or the advanced flexible prototypes meant for Mars, are simple in their basic purpose. They provide a layer of air around its occupant, using a body-shaped bladder that contains a simulated atmosphere. Besides delivering breathable air, space suits regulate temperature at a comfortable level and shelter the body against vacuum. Finally, it shields the wearer against the ravages of radiation, as well as micrometeoroids that can travel at a speedy 22,500 mph.

To survive these punitive conditions the space suits of the Apollo era had to evolve at an unbelievable pace to meet increasing demands of greater and greater heights and more and more dangerous environments, where even the most random movement might mean injury or death.

It was only a decade before the moon landing in 1969 that humanity broke through the 100,000-foot barrier for the first time, when Capt. Joe

Jordan flew his Lockheed F-104 Starfighter turbojet to an altitude of 103,389 feet. Given that President Kennedy, addressing a joint session of Congress on May 25, 1961, publicly declared "[T]his nation should commit itself to achieving the goal, before this decade is out, of landing a man on the moon and returning him safely to the earth," the learning curve for space suits was going to be as short as it was intense.

NASA convened the first space suit conference on January 29, 1959. The audience of more than forty experts advised the agency that there should be a far-reaching and exhaustive assessment program for what was uncharted territory. In the months following the symposium, three principal contestants emerged from the pack: the B. F. Goodrich Co. which had supplied most of the navy's pressure suits; the David Clark Co., which had built the air force's pressure suits; and the International Latex Corporation (ILC), a longtime bidder on government contracts involving rubberized material, of which NASA expected there was going to be a great deal in the astronauts' ensembles. After weighing the companies' various merits and approaches, NASA awarded Goodrich the prime contract for the Mercury space suit on July 22, 1959. One of the most senior designers from Goodrich was Russell Colley, whose bona fides included designing an aluminum helmet that helped aviator Wiley Post fly his aircraft *Winnie Mae* above the record 47,000-foot altitude.

Colley and Goodrich would base much of the Mercury apparel, such as the helmet, gloves, and coveralls for the torso, on the company's Navy Mark IV pressure suits. All the same, it would be forced to modify the suit extensively to meet the Mercury astronauts' special needs. Of these wide-ranging alterations, the most critical would be the air-circulation systems. The Mercury headpiece needed to be a "closed" system that operated on one air source for ventilation and breathing. This design allowed for a certain straightforwardness, augmented by the minimized weight of the ventilation/respiration equipment. NASA also lined the interior with a layer of neoprene-coated nylon and coated it with alumi-nized nylon on the outside to maintain the suit's internal temperature.

When they were finished, the bespoke suits looked like the Silver Surfer's hand-me-downs. In all, NASA ordered thirteen of the suits for astronauts Wally Schirra and John Glenn, and other space pilots, engineers, and testers. NASA followed up later with an additional order of eight more suits.

The pioneering astronauts of the Mercury program didn't have to move around much. They couldn't have even if they had wanted to, wedged in the cramped confines of the 6-foot-10-inch-long, 6-foot-2.5-inch diameter capsules. Nonetheless, the suits were so stiff and inflexible that being inside one was likened to living in a pneumatic tire. Fabric break lines sewn into the suits' joints offered a rudimentary mobility at the elbow and knee. Even then, once the suit was fully pressurized, an astronaut could barely bend his elbows or knee joints. As a result, the suit was worn "soft"—that is, unpressurized.

For all the effort that went into its development, the Mercury suit was worn by just half a dozen astronauts before it was replaced by the Gemini version. Designed by David Clark, the Gemini suits by necessity had to be more comfortable because the length of the Gemini missions would often far exceed Mercury's—the program's nineteen launches would include ten crewed missions, lasting from five hours to fourteen hours. The emphasis on mobility and range of motion continued with Gemini space suit designers nixing the Mercury suits' fabric-type joints and replacing it with a suit whose pressure bladder and "link-net restraint layer" made the suit flexible when pressurized.

Making the suit more comfortable sounds like a wimpy retreat from the manly manliness considered an indispensable quality of astronauts. But the importance of keeping an astronaut relaxed was reinforced by a report published in 2018 in the journal *Nature*. Researchers had pored over the historical biomedical data from Project Mercury and found that all the astronauts experienced heart-rate increases and weight loss just from being in the suits themselves.

The next step in space-suit fashion was the Gemini spacewalk suit. Known as the G4C, it was the first suit specifically designed to ramble outside the capsule. To survive the vacuum, the suit—which weighed as much as 34 pounds—was attached to the spacecraft with a hose that delivered oxygen to the spacewalking astronaut. Anticipating the worst-case scenarios of trouble getting back into the ship (like the near-disastrous

The cramped Mercury capsule. NASA

distress Alexey Leonov experienced on the world's first spacewalk), some alternative versions of the suit provided up to thirty minutes of backup life support. What made building this particular suit especially challenging was that it would be first used to venture outside the Gemini spacecraft. That meant an astronaut would dive into a firing squad of millions of

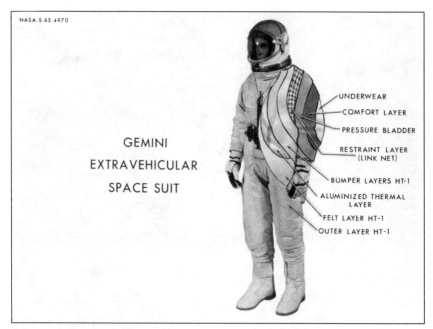

NASA.S.65.4970

GEMINI
EXTRAVEHICULAR
SPACE SUIT

UNDERWEAR
COMFORT LAYER
PRESSURE BLADDER
RESTRAINT LAYER
(LINK NET)
BUMPER LAYERS HT-1
ALUMINIZED THERMAL
LAYER
FELT LAYER HT-1
OUTER LAYER HT-1

Cutaway of Gemini extravehicular suit. NASA

microscopic bullets whose size is offset by their ability to essentially sand-blast away a surface. To protect the spacewalkers, layers of Dacron, Mylar, and other polymer materials were added to the suit to shield and insulate them, including Ed White, the first American to walk in space.

Other changes made later on reflected the unique problems posed by spacewalks, such as covering the metal zipper in back with a fabric flap, because if an astronaut turned his back to the sun, the zipper would quickly heat up.

When Neil Armstrong and Buzz Aldrin shuffled over the lunar surface, in Michelin Man–like enclosures, moving with the grace of Lurch hobbled by a Denver Boot, it might have been hard to believe those iconic protective suits had more in common with the Cross-Your-Heart-Bra than Cape Canaveral.

The challenge of the task was almost exponentially tougher than for Mercury or Gemini. It was analogous to going from scaling a climbing wall at the mall to ascending K2. The Apollo suits were slated to be the first that would shelter astronauts on extended walks of several hours on the moon's hostile surface. They had to be armored against a massive temperature flux of 540 degrees Fahrenheit (+260 degrees Fahrenheit to -280 degrees Fahrenheit), and regolith (dust and fragmented rocks) that can be as sharp as the kind of broken bottle you'd use in bar fight. The Apollo suit also had to be flexible enough to allow the astronaut to install gear and pick up moon rocks without undue strain. (One account likened the necessary elasticity to a "full range of body-joint motion as in the nude condition.")

The contest to win the contract for the Apollo suits began barely after the first blastoff in the space race. NASA requested that bids for the Apollo lunar suits be submitted by December 1, 1961, just a handful of months after Freedom 7, the first suborbital flight to put an American in space, went up. Eight companies submitted proposals, and in March of 1962, NASA selected Hamilton Standard, which would contract much of the work to International Latex Corporation (ILC), better known for their Playtex brand, and its Living Girdle, Living Bra, and later on, the Cross-Your-Heart Bra.

Hamilton Standard delivered a prototype suit called "Tiger" to NASA. It failed the agency's demanding tests badly. Although the deficient suit was more of a Hamilton Standard product than an ILC one, the company proved adroit at the blame game, pointing an accusatory figure at its subcontractor. Three years later, with no satisfactory suit in the offing, NASA invited Hamilton Standard and B. F. Goodrich to try again. ILC, armed with more pluck than power, fought for and won another chance to compete.

Managed by a former TV repairman and mechanic/pilot who'd sold sewing machines, the ILC troop of seamstresses and mechanics worked at times in twenty-four-hour shifts for six weeks—even picking the locks on the recently emptied Hamilton Standard offices in their facility to steal back their own supplies and records. When they were done, their suit won Best in Show, taking first place in twelve of twenty-two tests. The David had vanquished the larger, richer, and better-connected Goliaths, whose blunders included an exploding helmet and a suit too fat to fit through an Apollo capsule door.

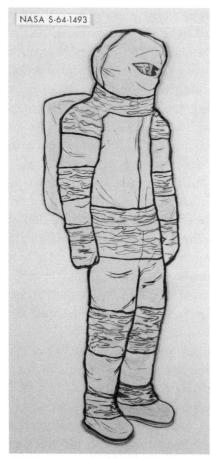

NASA S-64-1493

Artist's rendering of Apollo prototype suit. NASA

By 1968, ILC had produced the final ready-to-go lunar version, the 180-pound A7L suit. The total weight included a primary life-support system (when it was attached). On the moon, with a gravity one-sixth that of Earth, the outfit would feel like a comparatively featherweight 30 pounds. Top to bottom, helmet to boot, the A7L was a model of tailor-made ingenuity and individual craftsmanship.

The Apollo polycarbonate helmets were attached to the space suit by a pressure-sealing neck ring. Fixed into place, the helmet permitted the crew to swivel their heads freely inside—the opposite of the Mercury and Gemini helmets, which moved with the crewman's head, stressing with the extra weight. The Apollo helmet came with a feed port in case a crisis befell an astronaut, forcing him to stay in the suit longer than anticipated. To feed himself—or, if he was incapacitated, to allow his partner to feed him—he could introduce a tube filled with a John Glenn–era paste into the port.

Enclosed in multiple insulating layers, the gloves had been manufactured after taking individual casts of Neil Armstrong's, Buzz Aldrin's, and Michael Collins's hands to ensure a perfect fit. The thumb and fingertips were crafted out of a silicone rubber that would grant their hands a degree of sensitivity and "feel."

The Apollo suit consisted of a geologic strata–like twenty-one layers. Its outer cover—sometimes called the "Integrated Thermal Micrometeoroid Garment," or ITMG—was an assembly of multiple tiers designed to act as a shield against fire, heat, cold, radiation, abrasion, and, as its name makes obvious, micrometeoroid impacts.

Deeper down was Super Beta Cloth, a woven fireproof fiberglass that would protect the wearer against temperatures up to +1,200 degrees Fahrenheit. It was necessary insurance against a repeat of the Apollo 1 deaths that never left NASA's collective memory and had shadowed every launch since.

Further on was a bundle of eleven layers, including Mylar film, Dacron, and Beta Marquisette that also served as a supplementary radiation shield and bonus blockade against the micrometeoroids. The cosmic bullets would strike the ITMG's surface, then lose more and more speed and power as the smaller particles knifed through the multiple, tough protective layers, eventually losing all momentum. Lying flush against the astronauts' legs and waists was a pair of "long johns." Cooled water flowed through thin strips of tubing sewed into the old-timey-type underwear to convey body heat away.

Last, the lunar boot Armstrong took that "giant leap for mankind" in was a sort of overshoe pulled over the space suit's integrated pressure boot. Made mostly from a fabric of woven stainless steel, it had a tongue of Teflon-coated glass-fiber cloth and a sole of ribbed silicone rubber designed to handle the rough and spikey lunar surface.

The suits that protected the Apollo astronauts so well against squalls of heat and cold, blasts of radiation, and swarms of micrometeoroids are fading. They were built to withstand anything but time.

After the moon landing in 1969, NASA took Armstrong's suit on the road, a victory lap to the capitals of all fifty states and Washington, DC. Afterward it went on display until 2006, when curators at the National Air and Space Museum took it down. All those many layers of materials were decaying, with their off-gassing accelerating the degeneration.

The A7L was put to rest in a coffin-like enclosure where the temperature was always +60 degrees Fahrenheit and the humidity a constant

30 percent. Later, the conservators moved Armstrong's suit to a walk-in cooler at the Steven F. Udvar-Hazy Center, where it joined almost 270 other fellow space suits in storage. Now, thanks to its first Kickstarter campaign, "Reboot the Suit," that took place in 2015, the Smithsonian will be able to preserve the suit. The $719,779 raised from 9,477 backers will allow the museum to display Armstrong's suit in time for the fiftieth anniversary of the moon landing. Viewers will be able to view the historic garb up close—right down to the grains of lunar dust that rooted themselves in the suit from the knees down on that historic day, so many years and so many worlds ago.

CHAPTER TWENTY

Look to the Skies: UFOs

UNIDENTIFIED FLYING OBJECTS (UFOS) HAVE ALWAYS OCCUPIED THE left-hand side of the space program. In the early days, they were a shadow, a reflection, a distorted funhouse mirror image of NASA's efforts to venture into space that could never completely be explained away to anyone's satisfaction. They played the Joker to NASA's Batman, the anarchic Mr. Hyde to its upright Dr. Jekyll.

Some believed UFOs were harmless marsh gas. Others thought they hailed from far-away and much-advanced planets and gave us, accidentally or intentionally, the technology to travel off-world. No achievement in space travel could be viewed by this last group without a skeptical squint at NASA's low-ambition claims that the crew of Apollo 11 moseyed around on the moon for three hours, and did nothing more than pick up moon rocks, plant the US flag, and leave a plaque asserting their visit was of peaceful intent, blah, blah, official story, blah.

What really happened, according to a "suppressed" transcript that moves around the Internet, is a satellite of paranoia stuck in perpetual orbit: The stifled record picks up after Neil Armstrong and Buzz Aldrin exited the lunar module, Eagle.

Apollo 11: "Those are giant things. No, no, no, this is not an optical illusion. No one is going to believe this."

NASA: "What . . . what . . . what? What the hell is happening? What's wrong with you?"

Apollo 11: "They're here, under the surface."

NASA: "What's there? [muffled noise] Emission interrupted; inter-ference control, calling Apollo 11."

Apollo 11: "We saw some visitors. They were here for a while, observ-ing the instruments."

NASA: "Repeat your last information."

Apollo 11: "I say that there were other spaceships. They're lined up on the other side of the crater."

Apollo 11: "Let us sound this orbita . . . in 625 to 5 . . . automatic relay connected . . . my hands are shaking so badly I can't do anything. Film it? God, if these damned cameras have picked up anything—what then?"

NASA: "Have you picked up anything?"

Apollo 11: "I didn't have any film at hand. Three shots of the saucers, or whatever they were that were ruining the film."

NASA: "Control, Control here. Are you on your way? What is the uproar with the UFOs, over?"

Apollo 11: "They've landed there. There they are, and they're watch-ing us."

NASA: "The mirrors, the mirrors—have you set them up?"

Apollo 11: "Yes, they're in the right place. But whoever made those spaceships surely can come tomorrow and remove them. Over and out."

Even today, years after the infamous Roswell incident, and the UFO craze of the 1950s and '60s, the belief we are being visited by beings from another world remains rock steady, mulishly resistant to time and denial. Sure, after evaluating all UFO sightings from 2000 to 2014, *The Economist* magazine found that most people reporting the elusive aircraft had a blood-alcohol level of "plastered." Peter Sturrock, a professor of space science and astrophysics at Stanford University, surveyed members of the American Astronomical Society in 1977. He discovered that 4.6 percent of them—likely not inebriated—witnessed aerial phenomena beyond their ability to explain. A 1990 Gallup poll found that 27 percent of Americans believed "extraterrestrial beings have visited Earth at some time in the past." That number jumped up to 33 percent in 2001 and kept growing. When 20th Century Fox Home Entertainment commissioned a survey to promote the 2017 Blu-ray release of the movie about a mysterious UFO sighting, *Phoenix Forgotten*, it found nearly half of Americans reported they believe in the existence of aliens. Moreover, the survey found nearly as many believed aliens are visiting Earth, though fewer than 20 percent have confidence in the accounts of alien abduction, and a bit less than that claim to have seen a UFO.

The most publicly known of the US government programs that investigated this phenomenon was Project Blue Book. Run by the US Air Force, Project Blue Book operated from 1952 to 1969 (preceded by Project Sign and Project Grudge, which ran from 1947 to 1949, and 1949 to 1952, respectively). It was an era that spanned administrations from Dwight Eisenhower to Richard Nixon, music from Rosemary Clooney to Zager and Evans, and cinema from *Singin' in the Rain* to *Easy Rider*. The culture may have shook, rattled, and rolled in that interval, but the UFOs and our reaction to them were as constant as a metronome.

Agents for Project Blue Book researched 12,618 reports of UFOs. One portion of the report encompassed 37 cubic feet of case files holding records on individual sightings amounting to about 74,000 pages of files. Of all those cases they examined 701, or about 6 percent, remain officially unidentified.

It wasn't just the unwashed herd and huddled masses who believed flying saucers were real. President Jimmy Carter believed in them. Astronaut Gordon Cooper believed in them. Stephen Hawking believed in them.

Hillary Clinton in 2016 said she wanted to appraise files about UFOs and Nevada's legendary Area 51 site to see if the truth was in there. "I would like us to go into those files and hopefully make as much of that public as possible," she said. "If there's nothing there, let's tell people there's nothing there."

The belief, a phenomenon unique to the second half of the American century, even has its own sacred men and holy sites, its Moses, Mount Nebo, and Stonehenge, in Kenneth Arnold, Roswell, and Area 51.

UFOs have an origin story we know by heart as much as Superman's, with its set pieces of a doomed Krypton, desperate parents, and a tiny rocket that whisks a swaddled super-baby to the lonely Kent family in corn-filled Kansas.

While flying over Mount Rainier on June 24, 1947, Kenneth Arnold was flying his CallAir A-2, over Mineral, Washington, near Mount Rainer. Arnold was on the lookout for a Marine Corps C-46 transport airplane that had apparently crashed somewhere in the surrounding wilderness passing by beneath him. Whoever found the wreck would be in line for a $5,000 reward. His eyes were sharp and searching.

All of sudden, there was an intense, penetrating flash. Then, nine flashes in a row, fast as snapping fingers, that resolved into nine disk-like objects that the media later labeled "flying saucers." They moved in smooth unison, Arnold said of the soaring entities, like "the tail of a Chinese kite." Faster than you can say *Klaatu barada nikto*, the idea that we were being visited from other worlds spread with the speediness of memes and the doggedness of flu: One study of 149 newspapers that year found evidence of 853 flying-saucer sightings.

Not long after Arnold's encounter, W. W. Brazel discovered some strange fragments by his ranch near Roswell, New Mexico. His find included sticks, tinfoil, and strips of rubber. A few weeks later, Brazel, who did not own a telephone, turned the mystifying debris over to Roswell sheriff George Wilcox, who subsequently contacted the Roswell Army Air Field (RAAF). (The army and air force were still a combined service in 1947.) After military personnel collected the items, the RAAF on July 8, 1947, issued a press release that weirdly specified a "flying disk" had been recovered from a local ranch.

The local newspaper, the *Roswell Daily Record*, met with Lt. Walter Haut, author of the press release, ran the story, and then proceeded to give its own summary, reiterating much of what the RAAF had stated.

RAAF Captures Flying Saucer on Ranch in Roswell Region
No Details of Flying Disk Are Revealed

The intelligence office of the 509th Bombardment group at Roswell Army Air Field announced at noon today, that the field has come into possession of a flying saucer.

According to information released by the department, over authority of Maj. J. A. Marcel, intelligence officer, the disk was recovered on a ranch in the Roswell vicinity, after an unidentified rancher had notified Sheriff Geo. Wilcox here, that he had found the instrument on his premises.

Major Marcel and a detail from his department went to the ranch and recovered the disk, it was stated.

After the intelligence officer here had inspected the instrument, it was flown to higher headquarters.

The newspaper's account also included the additional information that a Mr. and Mrs. Dan Wilmot had spied the flying disk from their porch about ten p.m. An outsized object rocketed out of the sky from the southeast, the Wilmots related, glowing like a distant lightbulb that sped toward the northwest. Before July 8 was over, officials from the Army Air Force base in Fort Worth, Texas, had examined the wreckage. It was not a flying saucer, they said, but a high-altitude weather balloon equipped with a radar target made of aluminum and balsa wood.

Journalists were shown the foil, rubber, and wood rubble purportedly acquired from the crash area, which indeed looked like the kind of scrap a crash-landed weather balloon would leave behind. The *Roswell Daily Record* later ran a correction that included the military's statement that it was a weather balloon that had been found at the site.

Or maybe that's what the government wanted the happily gullible sheeple of 1947 to think. By the end of the year, what would eventually be crystallized as "the Roswell incident" had shuffled quietly into oblivion, a curiosity forgettable because it fit no established narrative in an era when the politicians and institutions still possessed a golden aura of "I cannot tell a lie." It was destined to be no more than a footnote in the annals of UFO literature until 1978, when Stanton Friedman, former nuclear physicist and UFO-ologist, interviewed Jesse Marcel, a retired member of the air force. Marcel claimed he had handled the wreckage of a downed spaceship at Roswell, though he couldn't exactly remember either the month, or even the year, the event took place. Even if it was vague and elusive, his testimony was enough to convince Freidman that there had been a cover-up of a "cosmic Watergate." His belief became the basis of a 1980 book called *The Roswell Incident*, co-written by Charles Berlitz (who authored *The Bermuda Triangle*) and William Moore. Roswell was back on the map.

Area 51, located about 80 miles northwest of Las Vegas, near Groom Lake and the northeast corner of the Nevada Proving Grounds, received its no-frills name from its numerical designation on a map.

The restricted military outpost, part of the Edwards Air Force Base, has been the site of numerous hush-hush military programs, and more myths than Mount Olympus. In the summer of 1955, observers began sighting UFOs around Area 51. Mainstream history would say that's because the air force had begun its testing of the recently developed U-2 aircraft then. With the standard airliner's 10,000- to 20,000-foot range, and military aircraft topping out around 40,000 feet, pilots who saw a zooming speck high above them at 60,000 feet might be understandably prone to concluding the streaking objects were something out of the science-fiction movies of the time, like 1951's *The Day the Earth Stood Still*.

In addition to the U-2 frequently taking off from Area 51, the CIA's A-12 OXCART and the air force's SR-71, both of which looked like the offspring of a black mamba and a stiletto, operated from there as well. The A-12 OXCART tore through the air at 2,208 mph, reaching 90,000 feet, while the SR-71 hit 2,193 mph, climbing to 85,069 feet. Add in another clandestine aircraft from years later, the Bird of Prey—so named because of its similarity to the Klingon vessel from *Star Trek*—and you have the raw ingredients for more UFO theories than *The Arabian Nights* has genies.

Instead of admitting they were testing top-secret advanced spy planes, the air force defaulted to stilted explanations that included "natural phenomena" and "high-altitude weather research," which only made UFOs seem like a more-reasonable alternative.

All of these explanations, and the reasonable denial embedded within, were meant to be a cool stream hosing down an out-of-control wildfire of speculation and rumor. In 1994, the air force, in response to a request initiated by New Mexico congressman Steven Schiff, looked into the Roswell incident. Its findings divulged that the material found at Roswell was from a US spy balloon, which had been a component of Project Mogul, designed to monitor the Soviet Union's nuclear tests. Three years later, the air force compiled a 1,000-page record of the report published in one volume, named *The Roswell Report: Case Closed*. It suggested that stories of alien bodies stemmed from civilian witnesses who saw crash-test dummies, made up to look like severely injured parachutists, as well as charred bodies from an airplane crash.

The most-read item on the CIA's website in 2014 was a report written in 1998. Titled "The CIA and the U-2 Program, 1954–1974," it has an entire section devoted to "U-2s, UFOs, and Operation Blue Book." The excerpts from that particular segment read with a tension between revelations about military secrets they were willing to make, and admissions about UFOs that skeptics might believe will likely never come.

> *High-altitude testing of the U-2 soon led to an unexpected side effect—a tremendous increase in reports of unidentified flying objects (UFOs). In the mid-1950s, most commercial airliners flew at altitudes between 10,000 and 20,000 feet . . . Consequently, once U-2s started flying at altitudes above 60,000 feet, air-traffic controllers began receiving increasing numbers of UFO reports.*

Later the report issues an RIP for all those who believed Earth was a frequent layover for alien itinerants:

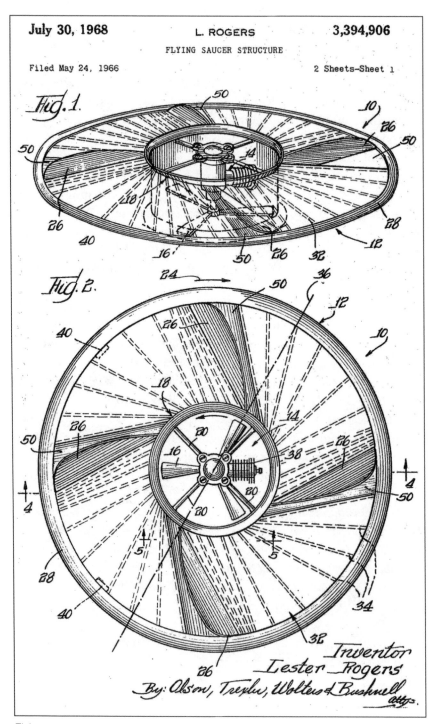

Flying saucer patent. USPTO

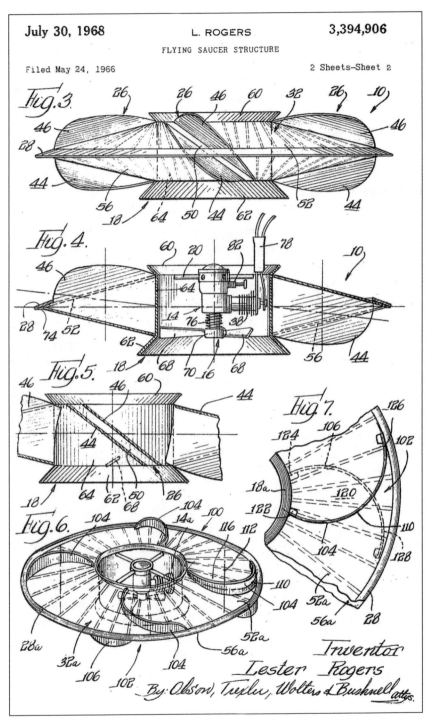

Flying saucer patent (*continued*)

[The U-2] flights accounted for more than one-half of all UFO reports during the late 1950s and most of the 1960s.

The subtext of all official explanation/disavowal is that when it comes to UFOs, people most believe what they most hope to be true. Like those who remain firm in their conviction that the 1969 moon landing was filmed in an Area 51 hangar. It's not just Internet trolls and conspiracy buffs who live in a van down by the river who snub the official story. Even those who were closest to the government brush it off with an equal or greater vigor. J. Allen Hynek, the American astronomer who consulted with Project Sign, Project Grudge, and Project Blue Book, came to believe there was an intricate government cover-up straight out of the 1970s paranoid cinema of *The Conversation* and *Three Days of the Condor*. Hynek (who invented the *Close Encounter* classification system for when we cross paths with ETs) said, "I remember the conversations around the conference table in which it was suggested that Walt Disney or some educational cartoon producer be enlisted in [the] debunking process."

According to their 2009 study, "Extraterrestrial Beliefs Scale," by psychologist Viren Swami of the University of Westminster and Malaysia University, along with University of Vienna psychologists Jakob Pietschnig, Stefan Stieger, and Martin Voracek, UFO believers scored high on what the researchers labeled "Openness to Experience." This meant they ranked high on such qualities as willingness to fantasize, explore new ideas, and try out new things.

Perhaps unspoken in that assessment is that believers in UFOs might presuppose spacecraft with the capacity to cross distances between stars at warp speeds would also have solved the daunting problems of pollution, warfare, and inequality. They had, like so many of us in the 1960s, grafted the optimism of *Star Trek* onto the slow but eventually-we'll-get-there space program. The most fervent believers might trust that the mockery that goes with their belief may be no more valid than the *New York Times*'s famous 1920 dismissal of Robert Goddard and his ridiculous ideas about space travel. They work under the radar, again, like Goddard perhaps, human engines of curiosity and optimism. Nowhere is that better evidenced than the United States Patent Office (USPTO),

US 20030122033A1

(19) **United States**

(12) **Patent Application Publication** (10) Pub. No.: **US 2003/0122033 A1**
Gao (43) Pub. Date: **Jul. 3, 2003**

(54) **RING-SHAPED WING HELICOPTER**

(76) Inventor: **Hengwei Gao**, Beijing (CN)

Correspondence Address:
PALMER & DODGE, LLP
KATHLEEN M. WILLIAMS
111 HUNTINGTON AVENUE
BOSTON, MA 02199 (US)

(21) Appl. No.: **10/295,796**

(22) Filed: **Nov. 15, 2002**

Related U.S. Application Data

(63) Continuation of application No. PCT/CN01/00803, filed on May 17, 2001.

(30) **Foreign Application Priority Data**

May 17, 2000 (CN) 00234123.9

Publication Classification

(51) **Int. Cl.^{7}** **B64C 15/00**; B64C 29/00
(52) **U.S. Cl.** **244/12.2**

(57) **ABSTRACT**

The invention discloses a new type ring-shaped wing helicopter, which is similar to a flying saucer in appearance with lift and attitude control torque brought by the blades of two ring-shaped wings rotating in opposite directions between the inlet cascade and the outlet cascade of the circular fringe of wing fuselage.

The helicopter of this kind without any outer rotors has high flying speed, good hydrodynamic form and simple-small structure and extensive securities—the crewman can easy escape quickly by ejecting upwardly in danger, the body of the helicopter allows slight impact with other objects during routine flight, and the running parts with high kinetic energy cannot threaten the personnel inside and outside the helicopter directly in its taking off and landing processes.

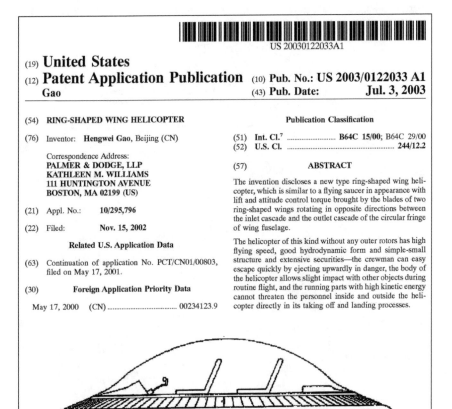

Ring-shaped helicopter / UFO hybrid. USPTO

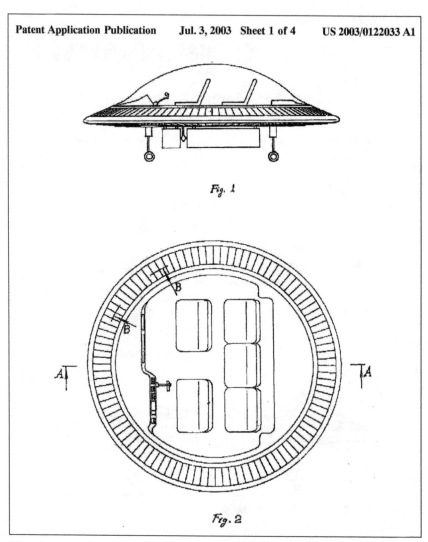

Fig. 1

Fig. 2

Ring-shaped helicopter / UFO hybrid (*continued*)

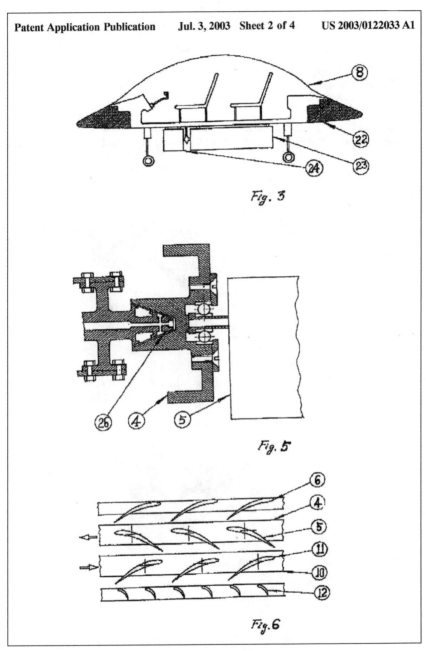

Ring-shaped helicopter / UFO hybrid (*continued*)

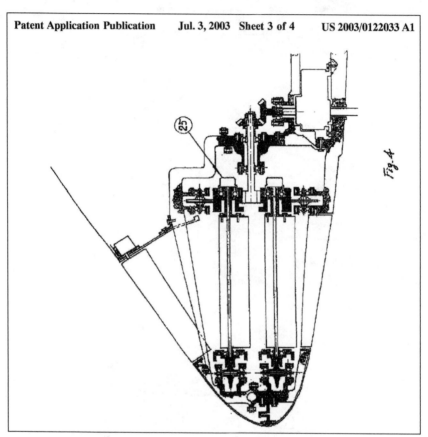

Ring-shaped helicopter / UFO hybrid (*continued*)

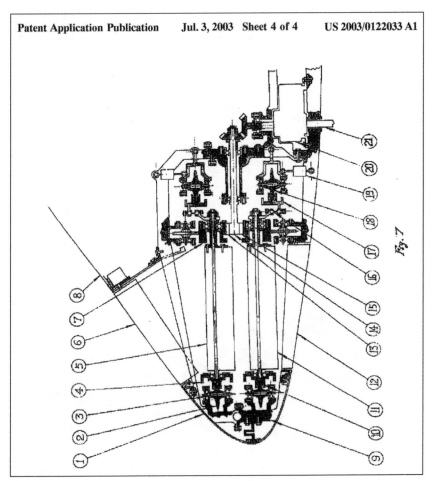

Fig. 7

Ring-shaped helicopter / UFO hybrid (*continued*)

(19) **United States**

(12) **Patent Application Publication** (10) Pub. No.: **US 2006/0145019 A1**

St. Clair (43) Pub. Date: **Jul. 6, 2006**

(54) **TRIANGULAR SPACECRAFT**

(76) Inventor: **John Quincy St. Clair**, San Juan, PR (US)

Correspondence Address:
JOHN ST. CLAIR
52 KINGS COURT, 4A
SAN JUAN, PR 00911 (US)

(21) Appl. No.: **11/017,093**

(22) Filed: **Dec. 20, 2004**

Publication Classification

(51) **Int. Cl.**
B64G 1/40 (2006.01)
(52) **U.S. Cl.** **244/171.5**

(57) **ABSTRACT**

A spacecraft having a triangular hull with vertical electrostatic line charges on each corner that produce a horizontal electric field parallel to the sides of the hull. This field, interacting with a plane wave emitted by antennas on the side of the hull, generates a force per volume combining both lift and propulsion.

Triangular spacecraft patent. USPTO

Figure 4

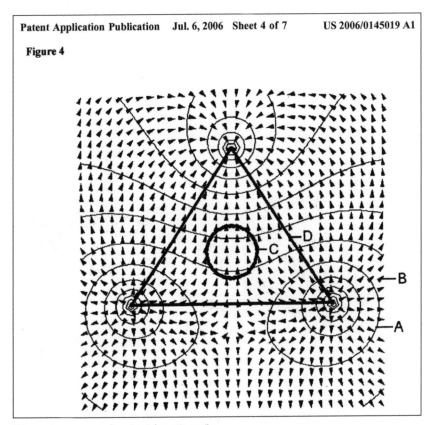

Triangular spacecraft patent (*continued*)

Figure 5

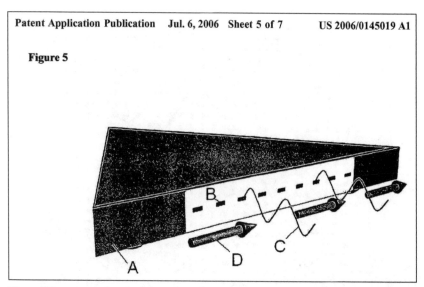

Triangular spacecraft patent (*continued*)

Figure 6

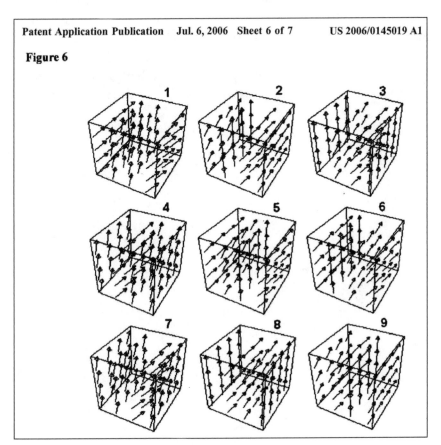

Triangular spacecraft patent (*continued*)

which even has a specific code for UFO-related inventions, B64C 39/001, which covers "flying vehicles characterised [*sic*] by sustainment without aerodynamic lift, often flying disks having a UFO-shape."

Nearly two hundred of the USPTO's filings fall under this classification, marked by curious spikes, especially between 1965 and 1971, the heyday of the Gemini and Apollo missions. Some claim these inventions are proof that aliens left us their scraps, as if from a picnic. Others would say they're just a meaningless mirror to their inventors' desires for the transformative technology of *Star Trek*. The fairest thing that might be said is that UFOs and the space race were both roads that led to unknown lands, and so will always have travelers ready to depart on them.

Index

About the Author

Charles Pappas has covered the expo industry for *Exhibitor* magazine since 2002, and is its de facto historian. Previously, he was the investigative reporter for Yahoo Internet Life, a columnist for Alexa.com, and a technology writer for *Small Office Computing*, *Home Office Computing*, and other publications. In the last few years his articles have won numerous national/regional ASBPE, MAGGIE, and TABBI awards. His previous books include *Flying Cars, Zombie Dogs, and Robot Overlords: How World's Fairs and Trade Expos Changed the World*, and *It's a Bitter Little World*, a revel in the tawdry neon language of film noir.